ACCIDENT PREVENTION

accident prevention

A WORKERS' EDUCATION MANUAL

INTERNATIONAL LABOUR OFFICE   GENEVA

HD
7262
A23
1983

ISBN 92-2-103392-9

First published 1961
Second edition 1983
Second impression (with modifications) 1986

The designations employed in ILO publications, which are in conformity with United Nations practice, and the presentation of material therein do not imply the expression of any opinion whatsoever on the part of the International Labour Office concerning the legal status of any country, area or territory or of its authorities, or concerning the delimitation of its frontiers.
The responsibility for opinions expressed in signed articles, studies and other contributions rests solely with their authors, and publication does not constitute an endorsement by the International Labour Office of the opinions expressed in them.
Reference to names of firms and commercial products and processes does not imply their endorsement by the International Labour Office, and any failure to mention a particular firm, commercial product or process is not a sign of disapproval.

ILO publications can be obtained through major booksellers or ILO local offices in many countries, or direct from ILO Publications, International Labour Office, CH-1211 Geneva 22, Switzerland. A catalogue or list of new publications will be sent free of charge from the above address.

Printed in Switzerland   IRL

PREFACE

In 1956, the ILO inaugurated a programme of workers' education with the object of giving workers a better understanding of the complexities of present-day social and industrial life and helping them to shoulder their responsibilities and safeguard their interests in the world of industry. The programme includes the organisation of courses and seminars, assistance to institutions responsible for workers' education, and the publication of the present series of manuals.

This manual deals with safety in industry, a subject of particularly direct interest to workers, for it concerns the preservation of life and limb.

It does not purport to deal comprehensively with the subject of occupational safety, the vastness and complexity of which are attested by the number of voluminous works written on each of its various aspects, such as safety in coal mining, in factories, in building and civil engineering, in agriculture, in forestry, in dock work, in the use of electricity, in the use of and maintenance of elevators or boilers, and other specialised subjects.

Nor does it pretend to cover every occupational field; it is, in fact, addressed almost exclusively to workers in manufacturing industries, and will give little specific guidance to the miner, the farm worker or the bricklayer.

Lastly, it is not a technical manual. It does not pretend to tell people how to prevent every kind of accident or guard every kind of equipment that could be met with in a factory. It only purports to explain why safety is important, by what methods it is promoted and what kinds of authorities, institutions and other organisations are responsible for promoting it.

The subject of occupational health is not touched upon here.

CONTENTS

THE PROBLEM OF ACCIDENTS DURING WORK

1

Every year, throughout the world, millions of industrial accidents occur. Some of them are fatal and some result in permanent disablement, complete or partial; the great majority cause only temporary disablement, which, however, may last for several months. Every accident causes suffering to the victim, a considerable proportion must cause much anguish to his or her family, and many — especially those resulting in death or permanent disablement — may have a catastrophic effect on family life. Moreover, all accidents waste time and money.

The world is still paying heavily for accidents in terms of both human suffering and economic waste. Despite some progress, the question of safety at work is still a serious problem.

Some idea of the size of the problem can be given by recalling that, during the six-year period of the Second World War, far more people were injured in accidents at work all over the world than were wounded as the result of hostile action. The figures given for the United Kingdom and the United States amply illustrate the point. Over the duration of the war, monthly casualties in the armed forces of the United Kingdom (excluding merchant seamen) averaged 3,462 killed, 752 missing and 3,912 wounded — a total of 8,126. During the six years from 1939 to 1944, in manufacturing industry alone (including docks and shipyards), the monthly average was 107 deaths and 22,002 injuries. In the United States armed forces during the Second World War, the average monthly losses were 6,084 killed, 763 missing and 15,161 injured, a total of 22,008; while the monthly average of industrial casualties during the years 1942-44 was 1,219 persons killed, 121 permanently and totally disabled, 7,051 permanently and partially disabled and 152,356 temporarily disabled — a total of 160,747.

It can thus be seen that in these two countries industrial injuries caused more casualties (leaving aside for the moment all questions of relative severity) than the operations of a major war.

Every year in the United Kingdom about 1,000 people are killed at their work. Half a million workers suffer various injuries, and 23 million working days are lost annually because of industrial injury and disease. In the United States, according to the National Safety Council, the frequency rate for disabling injuries (the number of reportable cases per million work-hours) rose from a low of 5.99 for 1961 to 10.87 for 1976, representing a huge increase of 81 per cent. The Council also estimated that accidents cost the nation US\$51,100 million in lost wages, medical expenses, damage to property and administration costs. In 1976 alone, a million productive work-years were lost through accidents at work. In everyday terms, these figures mean that injuries to workers would have had the same economic effect on the United States as if the whole of the nation's industry had been closed for one full week.

Today some countries (Japan, United States) regularly report over 2 million occupational accidents a year, and others (France, Federal Republic of Germany, Italy) over a million. Many countries, including some of the largest or most highly industrialised, still do not publish any figures; but it is fairly safe to assume that over 15 million occupational accidents occur throughout the world every year—a staggering number when considered in terms of the suffering, sorrow and waste they cause.

In a report on working conditions and environment submitted to the International Labour Conference in 1975, the Director-General of the International Labour Office gave an account of the situation with regard to occupational injuries in which it was stressed, in particular, that the accident frequency rates had stagnated in most of the industrialised countries and had risen in the developing countries.

The situation has not changed substantially since that time, as may be seen from the frequency rates of fatal occupational accidents in those countries for which comparable data covering several years are available. While some industrialised countries have succeeded in breaking through the levels of those frequency rates and in setting them on a slightly downward course, other industrialised countries have not yet been able to do so. The figures below illustrate the point.

The International Labour Office publishes comprehensive tables by individual member States of the annual incidence or frequency rates of fatal accidents for four industrial groups: manufacturing, construction, railways, and mining and quarrying. Published rates for some European Community countries and certain other major industrial countries are presented in the figures below. Differences in definitions of other accidents are thought to be so wide as to invalidate any significant international comparison, but the trends are evident. Also, the fatal accident rates quoted for the different countries vary in definition and in the nature of the information used to calculate them. The main distinction is between incidence rates, which relate deaths to number of wage earners or employees (or work-years) and frequency rates which relate deaths to work-hours.

2

All the incidence rates are broadly comparable with one another, as are all the frequency rates. In general, it is assumed that the numbers at risk are taken to include all employees, and known exceptions from this assumption are recorded in the footnotes; but, of course, the footnotes cannot cover all differences in definition. In addition, in such a broad sector as manufacturing, any differences in accident rates may merely reflect the fact that some countries have a larger proportion of high risk industries in this sector than others; there may be no real difference in the accident rates in individual industries.

Statistics such as those given above, when covering a sufficiently long period, certainly do give useful indications of general trends; but it would be difficult to draw valid inferences from those data, or to use them for making valid comparisons, without taking into account the circumstances

Figure 1.   Fatal accident rates in manufacturing industry for selected countries, 1973-77

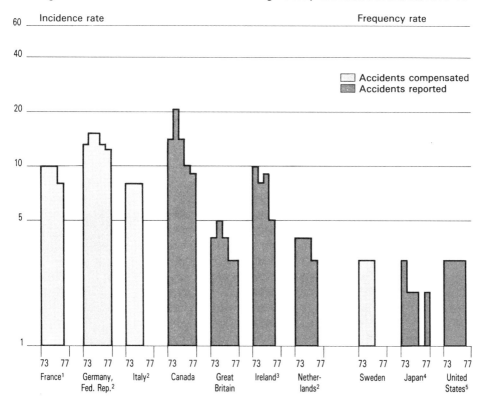

Incidence rate: deaths per 100,000 employees
Frequency rate: deaths per 100 million work-hours

[1] Including mining and quarrying.   [2] Based on standard work-years of 300 working days. With a five-day week, accidents per actual work-year are likely to be some 20 per cent fewer.   [3] Rate per 100,000 wage earners.   [4] Establishments employing 100 or more workers.   [5] Based on sample surveys.

Sources: London, Health and Safety Executive; Geneva, International Labour Office.

Accident prevention

in which the data were collected and compiled, and without adequate knowledge of their social, economic and cultural contexts. No single set of statistics can provide a truly comprehensive picture, as the criteria vary from country to country. But, provided the reference period is long

Figure 2. Fatal accident rates in the construction industry for selected countries, 1973-77

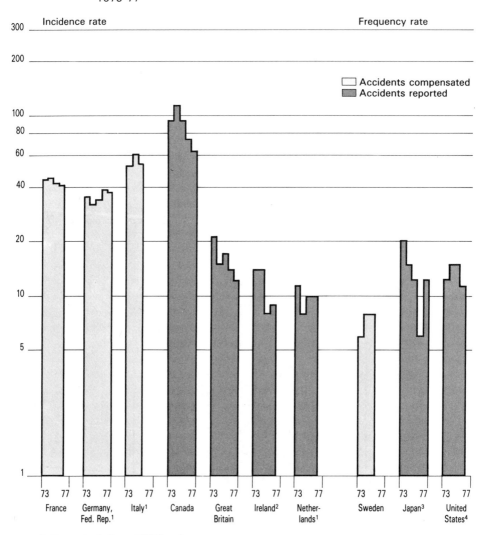

Incidence rate: deaths per 100,000 employees
Frequency rate: deaths per 100 million work-hours
[1] Based on standard work-years of 300 working days. With a five-day week, accidents per actual work-year are likely to be some 20 per cent fewer.   [2] Rate per 100,000 wage earners.   [3] Establishments employing 100 or more workers.   [4] Based on sample surveys.
Sources: London, Health and Safety Executive; Geneva, International Labour Office.

enough, inferences may legitimately be drawn from such statistical evidence.

However, as is often pointed out, statistics of occupational accidents diminish in practical value as one moves away from the plant level to the level of a sector of economic activity and to the national level. It is, of course, obvious that any approach in trying to prevent accidents requires adequate statistical information; but the lessons to be drawn from such information are usually confined in scope to the unit or to the department or, at most, to the undertaking concerned. It is only at the local level that those responsible for preventive measures can gauge correctly the influence on frequency and severity rates of the measures that they have taken (or not taken) and identify and evaluate with sufficient accuracy,

Figure 3.   Fatal accident rates on the railways for selected countries, 1973-77

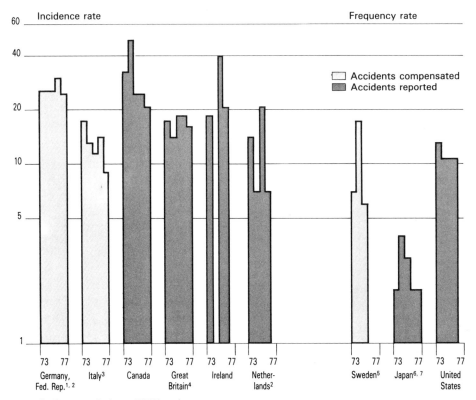

Incidence rate: deaths per 100,000 employees
Frequency rate: deaths per 100 million work-hours

[1] Including railway workshops and accidents involving road vehicles operated by federal railways.   [2] Based on standard work-years of 300 working days. With a five-day week, accidents per actual work-year are likely to be some 20 per cent fewer.   [3] Regular staff only; including railway workshops.   [4] Rate per 100,000 wage earners.   [5] Including railway workshops and construction of railway lines.   [6] Including railway workshops.   [7] Establishments employing 100 or more workers.

Sources: London, Health and Safety Executive; Geneva, International Labour Office.

5

both quantitatively and qualitatively, the various factors that may have influenced a particular trend.

So far, we have been considering mainly the industrialised countries; but nowhere is the problem of accidents more acute than in the developing

Figure 4. Fatal accident rates in the mining and quarrying industries for selected countries, 1973-77

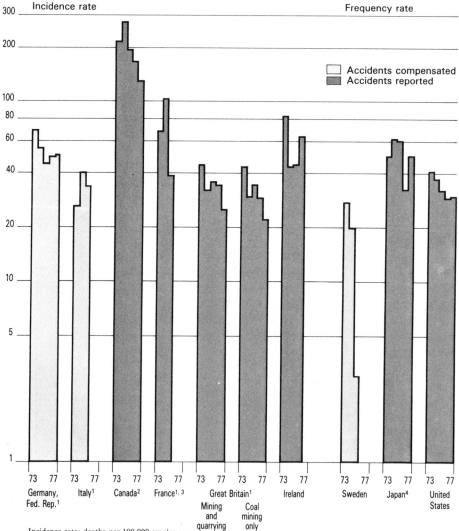

Incidence rate: deaths per 100,000 employees
Frequency rate: deaths per 100 million work-hours

[1] Based on standard work-years of 300 working days. With a five-day week, accidents per actual man-year are likely to be some 20 per cent fewer. [2] Rate per 100,000 wage earners. [3] Excluding quarrying. [4] Establishments employing 100 or more workers.

Sources: London, Health and Safety Executive; Geneva, International Labour Office.

countries. In the majority of these, the frequency rates of fatal occupational accidents, which are the only ones for which reliable data are available, show a marked upward trend; it is not unusual to find that the rates have doubled or tripled over the past ten or 15 years, or risen even more in those sectors of economic activity which have undergone a particularly rapid growth in the country concerned. The positive results that have been obtained in a very small number of developing countries emphasise all the more, by contrast, the failure of the others to reduce their accident rates. The situation is made far worse in these developing countries because protective laws may not exist; or if they do, they are rarely enforced. Also, where unemployment is endemic, the availability of large labour force reserves does not always result in improvements in safety. Poverty and ignorance also add to the problem.

This brief introduction to the size of the problem contains a number of statistics. Many would dispute them because a great number of accidents are never reported and many industrial diseases go unrecognised. However, as was stated earlier, one fact is indisputable — the world is paying a heavy price for accidents in terms of both human suffering and economic waste.

THE COST OF ACCIDENTS

Much has been written about the economic cost of industrial accidents but few attempts have been made to assess it accurately. It is important to question whether meaningful costs can actually be given to accidents, and, if so, what use can be made of them in terms of accident prevention. After all, what price can be put on a person's life?

The economics of accidents is inevitably linked to the economics of accident prevention. It is well known that damage and injury entail costs as does accident prevention. If the cost of accidents actually appears on a balance sheet, the employer, who is ultimately responsible, can integrate measures for accident prevention into the overall plans for the firm.

Various estimates have been made for the annual cost of industrial accidents in strictly monetary terms. In Great Britain, for example, Beckingsale in 1963 estimated the cost of industrial accidents to the nation at £264 million. In 1970, Hanna put the cost at £220 million. In *Safety and health at work*, the Report of the Committee under the Chairmanship of Lord Robens (1970-72), estimates ranging from £200 million to £900 million were put forward as the total cost of accidents to the nation. The report also stated that the total costs of accidents and industrial diseases in 1969 represented 0.87 per cent of the gross national product.

Figures exist for other countries and have been published in various texts over the past few years. For example, an American author estimated that in 1953 each lost-time injury (i.e. an injury involving absence from work for a certain time) in industrial employment in the United States cost the employer approximately US$1,800. According to the same author, the American Social Security Administration estimated that in a typical year it paid out compensation amounting to approximately US$535 million, while the National Safety Council estimated its medical expenses at US$130 million, making a total direct expenditure of US$665 million. The number of lost-time injuries which gave rise to this expenditure was 1.95 million; thus the average cost of each was US$340.

Other readily available statistics for the United States put the cost of accident injuries in 1965 at US$11,000 million, and that figure excludes material damage. As stated earlier, the National Safety Council estimated the cost of accidents at US$51,100 million in 1976.

Estimates for the cost of accidents for other countries were given at a colloquium on the prevention of occupational risks held by the International Social Security Association in 1965 and were published in *International colloques on the prevention of occupational risks: Colloque II* (Geneva, 1967).

But what of the 1980s? According to estimates recently made in some of the industrialised countries, the average overall costs of occupational accidents and diseases is equivalent to about 4 per cent of the gross national product of those countries. That represents a large increase over any previous estimates.

It is important to explain what precisely is meant by the total or overall cost of accidents because, whereas many of the costs of accidents can readily be expressed in monetary terms, others are less tangible. According to some authors, accident costs are made up of direct or subjective expenses comprising, for example, personal suffering and bereavement of the victim's family, and indirect, hidden or resource expenses which include material damage, loss of equipment, expenses resulting from loss of production and so on. There appears to be little agreement on what precisely constitutes an indirect cost, largely because so many variables are involved. As long ago as 1959, Heinrich listed the following hidden accident costs:

1. Cost of lost time of injured employee.
2. Cost of time lost by other employees who stop work:
    — Out of curiosity;
    — Out of sympathy;
    — To assist injured employee;
    — For other reasons.
3. Cost of time lost by foremen, supervisors, or other executives as follows:
    — Assisting injured employee;
    — Investigating the cause of the accident;
    — Arranging for the injured employee's production to be continued by some other employee;
    — Selecting, training, or breaking in a new employee to replace the injured employee;

  — Preparing state accident reports, or attending hearings before state officials.

4. Cost of time spent on the case by first-aid attendant and hospital department staff, when not paid for by the insurance carrier.

5. Cost due to damage to the machine, tools, or other property or due to the spoilage of material.

6. Incidental cost due to interference with production, failure to fulfil orders on time, loss of bonuses, payment of forfeits, and other similar causes.

7. Cost to employer under employee welfare and benefit systems.

8. Cost to employer of continuing the wages of the injured employee in full after his return, even though the services of the employee (who is not yet fully recovered) may be worth only about half of their normal value for a time.

9. Cost due to the loss of profit on the injured employee's productivity, and on idle machines.

10. Cost that occurs in consequence of the excitement or weakened morale due to the accident.

11. Overhead cost per injured employee — the expense of light, heat, rent, and other such items, which continues while the injured employee is a non-producer.

More recently, other authors have suggested that there may be as many as 26 factors involved in indirect costs.

Let us look at some estimates of the cost of industrial accidents in Great Britain based on two sets of figures taken from the Robens Report. These figures illustrate some of the differences between direct and indirect costs in strictly monetary terms. The figures in table 1 are based on information provided by Her Majesty's Factory Inspectorate (HMFI) and relate to employed persons killed or injured in accidents at premises covered only by the Factories Act. Those in table 2 are based on statistics from the Department of Health and Social Security (DHSS) and relate to accidents arising out of, and occurring in the course of, employment covered under the National Insurance (Industrial Injuries) Act.

Although these figures may not apply directly to other countries, they do indicate the way in which the component costs of accidents can be measured. They also allow a comparison or ratio to be drawn from the direct and indirect costs. In the HMFI data, this indirect/direct cost ratio is roughly 3 to 1. The ratio in the DHSS data is roughly 2 to 1. One of the first proponents of the use of this ratio, Heinrich, in 1959, calculated the ratio as 4 to 1. De Reamer, in 1980, cites ratios varying from 2.3 : 1 to 101 : 1.

Although it is now generally agreed that the indirect cost amounts to between two and five times the direct cost, depending on the circumstances, many authors have, nevertheless, questioned whether it is meaningful to compare such figures, as the ratios vary considerably from industry to industry depending on the type of injury, the damage caused and many other variables. Certainly it is important to put costs to accidents, but they must be real costs upon which positive action in terms of accident prevention can be based, rather than nebulous costs which

Accident prevention

Table 1. Estimates for the national costs of occupational accidents: HMFI data[1], 1969

Costs	Amounts (£)
Direct costs	**36 626 000**
Indirect costs	**97 291 226**
Fatalities	8 277 043
Industrial accidents	54 731 681
Under-reporting	19 774 952
Non-reportable accidents	14 507 550
Total	**133 917 226**

[1] Assuming damage and administration costs of £50 per accident.

Table 2. Estimates for the national costs of occupational accidents: DHSS data, 1969

Costs	Amount (£ million)	
	Conservative estimate	Best estimate
Direct costs	**110.59**	**127.42**
Fatalities	9.59	9.59
Serious injury	21.04	42.08
Slight injury	79.96	75.75
Indirect costs	**184.0**	**205.2**
Fatalities	20.2	20.2
Industrial accidents:		
loss of output	84.5	84.5
medical and hospital costs	8.1	12.4
Damage and administration costs	25.2	42.1
Long-term incapacity	8.1	8.1
Non-reportable accidents	37.9	37.9
Total	**294.59**	**332.62**

often cloud the issue and are open to dispute from all sides of industry. A good example of real costs is the cost of workers' compensation. There can be no doubt that increases in workers' compensation over the past few years have been an important factor in bringing to the attention of all concerned the urgent need for improvements in accident prevention measures. For example, the total cost of workers' compensation paid by

United States business firms rose from just under US$732 million in 1948 to over US$7,000 million in 1976.

The costs of accident prevention are relatively easy to assess in comparison with the costs of accidents. Sinclair divided accident prevention costs into three separate areas. Firstly, there are the design costs, such as those costs required to install machine guards to protect workers. Secondly, there are the operational costs, such as the costs of running a safety department (e.g. salaries, training, protective clothing, etc.). Lastly, there are the planning and consequence-limiting costs which refer to "safeguarding the future", as for example in the costs of environmental sampling or the cost of safety audits.

As was stated earlier, the economics of accidents is closely associated with the economics of accident prevention. As more money is spent on accident prevention, so the cost of accidents is reduced. However, there must come a point when, theoretically, more money could be spent on accident prevention than is saved in total accident costs. However, one is reminded of the question – what monetary value can be placed on a person's life and limb?

HOW ACCIDENTS ARE CAUSED

In every sphere of human activity there is the possibility of an accident, and work is no exception. Industrial accidents are the end-products of unsafe acts and unsafe conditions of work. However, accidents are preventable – they don't just happen. They usually occur as a result of the combination of a number of factors of which the three main ones are technical equipment, the working environment, and the worker. For example, safety equipment may be lacking in the factory or machinery may have been poorly designed with inadequate safety devices. The working environment may be so noisy that it is impossible to hear safety signals. The temperature may be such that workers become easily tired and are unable to concentrate on the task in hand, or inadequate ventilation may result in the build-up of toxic fumes and again lead to accidents. Also, the workers themselves may be a contributory factor in that they may not have received adequate training or may have little experience of the task. This is particularly relevant when new procedures are introduced into a factory or when people change jobs as, for example, in developing countries when workers leave the land to work in industry.

Ultimately, all industrial accidents are – either directly or indirectly – attributable to human failings. People are not machines; their performance is not fully predictable and mistakes are made. A mistake may be made by the architect who designed a factory, the contractor who built it, a machine-designer, a manager, an engineer, a chemist, an electrician, a supervisor, an operator, a person carrying out maintenance — in fact, by anyone who has anything to do with the design, construction, installation, management, supervision and use of the factory and anything in it.

Much thought has been given to the study of causes of accidents and many books have been written on the subject. In simple terms, if the causes of accidents can be found, the appropriate measures can be taken to prevent them. If the preventive measures are not taken, the same type of accident will occur again and again.

Accident causation is a complex subject and various theories have been put forward to explain how accidents happen and how they can be avoided in the future. For example, the "pure chance theory" implies that accidents are "acts of God", that there is no discernible pattern in the chain of events leading up to the accident, and that, as the name of the theory suggests, they depend entirely on chance. The "accident prone theory" suggests that some workers are more likely than others to have an accident due to innate personal characteristics. In other words, these workers are always likely to have an accident and very little can be done to prevent it. To ascribe the causation of an accident merely to "carelessness" on the part of a worker does nothing to identify the real cause. These and similar negative fatalistic theories have done little to advance the cause of accident prevention. They are easy excuses rather than root causes.

One method employed to throw some light on the actual causes of industrial accidents is classification. There are many different methods of classifying accidents according to causes; some are known as simple classification systems and others as multiple classification systems. Nearly every country has a different method. Some classify accidents according to where the fault lies, others classify them according to the cause. In some cases this has been done on the basis of a resolution adopted by the First International Conference of Labour Statisticians, organised by the ILO in 1923, which recommended a simple classification system of accidents by cause containing the following main headings: machinery, transport equipment; explosions and fire; poisonous, hot or corrosive substances; electricity; falls of persons; stepping on or striking against objects; falling objects; handling without machinery; hand tools; animals; and other causes.

This simple classification based on the cause of the accident proved unsatisfactory because of widespread interpretations of each cause, and because of the number of factors that, in combination, may result in an accident.

Accordingly, to overcome the objections, a number of countries and agencies within those countries adopted a more comprehensive multiple classification system in an attempt to identify under several possible headings the various factors involved in an accident. For example, the French Electricity Board *(Electricité de France)* has adopted a highly complex classification system in which each accident is tabulated according to a number of main criteria:

(a) the agency;

(b) the nature of the work;

(c) the defective condition of the agency;

(d) unsafe acts;

(e) psychological and human factors, climatic conditions, the conditions of the working surface, etc.

Another multiple classification system is that of the American National Standards Institute. Each essential factor of the accident falls into one of the following seven classes:

(a) nature of injury — identifies the injury in terms of its principal physical characteristics;

(b) part of the body affected — identifies the part of the injured person's body directly affected by the injury previously identified;

(c) source of injury — identifies the object, substance, exposure or bodily motion which directly produced or inflicted the previously identified injury;

(d) accident type — identifies the event which directly resulted in the injury;

(e) hazardous condition — identifies the hazardous physical condition or circumstance which permitted or occasioned the occurrence of the previously named accident type;

(f) agency of accident (or agency of accident part) — identifies the object, substance or premises in or about which the previously named hazardous condition existed;

(g) unsafe act — identifies the violation of a commonly accepted safe procedure which directly permitted or occasioned the occurrence of the previously named accident type.

In October 1962, the Tenth International Conference of Labour Statisticians, convened by the ILO, adopted a standard multiple classification system to replace the 1923 cause classification. According to this system, industrial accidents are to be classified under each of the following four headings: *(a)* the type of accident; *(b)* the agency; *(c)* the nature of the injury; and *(d)* the bodily location of the injury.

Whatever form of classification is adopted, it appears that the most common causes of accidents are to be found, not in the most dangerous machines (such as circular saws, spindle-moulding machines and power presses) or the most dangerous substances (such as explosives and volatile flammable liquids) but in quite ordinary actions like stumbling, falling, the faulty handling and lifting of goods or use of hand tools, and being struck by falling objects. In Great Britain in 1977, an analysis of factory (process) accidents revealed that 30 per cent of the total accidents occurred in handling goods, 16 per cent were falls and 14 per cent machinery accidents. Similarly, an analysis of construction (process) accidents for the same period indicated that nearly one-third of all accidents in building operations were caused by falls, while over one-quarter occurred in handling goods. At engineering construction works, the most common type of accident was in handling goods (over 25 per cent of the total), closely followed by falls (nearly 20 per cent). In 1977, on docks, wharves and quays, 27 per cent of the total number of accidents were due to falls, and 26 per cent occurred in handling goods. All these figures illustrate the everyday nature of accidents.

HOW ACCIDENTS ARE PREVENTED

The reader has already been introduced to the concept that accidents are the result of unsafe acts and unsafe conditions, and these in turn depend on a whole variety of factors. It is the interplay of these factors in a certain sequence that produces an accident. Any alteration in the sequence, or elimination of one of the factors in the accident chain, will usually prevent the accident. It is rather like dominoes which are placed on edge in a line next to each other. When the first falls, it automatically knocks down all the others; but if any domino is removed, the subsequent dominoes remain standing — that is, the accident will not take place. This implies that accident prevention is simply the removal of one of these factors. Unfortunately, just as causation is complex, so is accident prevention.

The various means generally used at present to promote industrial safety may be classified as follows:

(a) *regulations,* i.e. mandatory prescriptions concerning such matters as general working conditions, the design, construction, maintenance, inspection, testing and operation of industrial equipment, the duties of employers and workers, training, medical supervision, first aid, and medical examinations;

(b) *standardisation,* i.e. the laying down of official, semi-official or unofficial standards concerning, for example, the safe construction of certain types of industrial equipment, safe and hygienic practices, or personal protective devices;

(c) *inspection,* i.e. the enforcement of mandatory regulations;

(d) *technical research,* including such matters as investigation of the properties and characteristics of harmful materials, the study of machine guards, the testing of respiratory masks, the investigation of methods of preventing gas and dust explosions, or the search for

the most suitable materials and designs for hoisting ropes and other hoisting equipment;

(e) medical research, including, in particular, investigation of the physiological and pathological effects of environmental and technological factors, and the physical circumstances conducive to accidents;

(f) psychological research, i.e. investigation of the psychological patterns conducive to accidents;

(g) statistical research to ascertain what kinds of accidents occur, in what numbers, to what types of people, in what operations, and from what causes;

(h) education, involving the teaching of safety as a subject in engineering colleges, trade schools or apprenticeship courses;

(i) training, i.e. the practical instruction of workers, and especially new workers, in safety matters;

(j) persuasion, i.e. the employment of various methods of publicity and appeal to develop "safety-mindedness";

(k) insurance, i.e. the provision of financial incentives to promote accident prevention, in the form, for instance, of reductions in premiums payable by factories where safety measures of a high standard are taken; and

(l) safety measures within the individual undertaking.

It may be said that in the last resort the value of items *(a)* to *(k)* will depend very largely on the effectiveness of the last. It is in undertakings that accidents occur, and the accident pattern in a given undertaking may be determined very largely by the degree of safety-mindedness shown by all those who work in it.

It will be clear from this list that accident prevention requires the co-operation of many kinds of people — legislators, government officials, technologists, physicians, psychologists, statisticians, teachers and, of course, the employers and workers themselves.

ACCIDENTS AND ACCIDENT PREVENTION TODAY

The first great change for hundreds of years in the nature of industrial hazards came with the introduction of steam-powered machinery. Later came electricity, the use of which gave rise to yet other types of accidents. The pattern of risk has also been changed by the replacement of coal by gas and oil; the internal-combustion engine, too, has its dangers. The continual spread of mechanisation and the ever-increasing variety of industrial chemicals in use have added yet further problems of protection. More recent hazards are those deriving from ionising radiation and atomic power. Every day, new hazards are uncovered. It is all too easy to think that accidents are the natural and inevitable price of progress. However, despite the technology involved, the programme for putting men on the moon had one of the best accident records and shows what can be achieved. Thus, technological advances cannot be held solely responsible for the ever-growing number of accidents.

Technological changes do not always result in a net increase in the degree of risk. The individual drive for machines is undoubtedly safer than the old line shafting; the modern electric motor for cranes is safer than the old steam-engine; mechanical handling equipment prevents injuries due to over-exertion; and pneumatic conveyors prevent harmful dusts from entering the atmosphere.

Early attempts at preventing industrial accidents focused largely on machine guarding and safety devices, but it was soon realised that mechanical safeguarding alone was not enough and did little to eliminate the root causes of accidents. Gradually, the human element in accident prevention was recognised along with the need for safety education.

The war on accidents has developed from the tentative and sporadic measures of a century ago into a full-scale campaign in which almost every conceivable weapon, ranging from imposing codes of regulations down to cartoons, is being used. In this war, great successes have been achieved, but final victory — the reduction of accident frequency and severity rates to the lowest figure attainable by human effort — is still a long way off.

Although it may not always be possible to fix a definite goal for accident prevention, there is a wide measure of agreement that most accidents can be prevented and that the goal of a safe workplace should be aimed at. Companies with good safety organisation have actually proved that this is so in a large number of cases. If every undertaking in each industry were to reduce its accident rate to that attained by the few companies with the best safety records in the industry, without doubt the world accident total would be only a fraction of what it is now.

In the following pages an attempt is made to give a general idea of what accident prevention involves in the way of resources and effort, and to describe some of the tasks that have to be performed and the agencies established to perform them.

QUESTIONS

1. Do you consider that the prevention of occupational accidents is a serious problem? Give reasons for your answer.

2. Why is it difficult to explain satisfactorily how accidents happen?

3. Describe some of the ways in which endeavours are being made to prevent accidents.

4. Discuss whether technological progress is likely to reduce the danger of accidents in industry or not.

THE ORIGINS OF ACCIDENT PREVENTION

2

Industrial accidents first began to occur in large numbers over 150 years ago as the revolution in industrial techniques began to make possible large-scale mechanised production with the factory as the production unit. Some of the conditions to which the Industrial Revolution gave rise, as it ran its triumphant but pitiless course, were so atrocious as to create a widespread feeling of horror. The daily toll of accidents among workers inevitably gave rise to demands for reform, largely championed by the trade unions.

The movement for reform was led by people who felt that they had a moral responsibility for the well-being of their fellows. Accident prevention work has from the very beginning owed much to these public-spirited men and women, whose sense of justice was outraged by the exploitation of the weak and whose sympathy was stirred by their sufferings. The aim of the reformers was to persuade or shame the government into protecting the factory workers (and above all the children), who often lived and worked under conditions which today would be considered shocking, from the danger of mutilation, disease and immorality by taking, among other things, measures to reduce the frequency of industrial accidents.

If one takes as an example the country in which the Industrial Revolution began — the United Kingdom — one finds that these humanitarian efforts were first of all directed towards reducing the hours of work and protecting the health of children, who were by far the worst sufferers from these conditions, and that it was only at a comparatively late stage that any action was taken to prevent accidents in general.

In the eighteenth century, as a result of a remarkable series of inventions, the textile industry gradually developed from a cottage industry into a factory industry. Unfortunately, improvements in industrial safety were slow to follow, and much of the legislation, which at that time dealt largely with the textile industry, was rarely enforced. There arose a great demand for cheap labour, and a convenient supply was found among the pauper children who were in the care of the public assistance authorities of the large towns. They worked "unknown, unprotected and forgotten", as a writer described them in 1795, in unsanitary conditions, for up to 14 or 15 hours a day. During the next 40 or 50 years, as a result of more or less continuous agitation, much was done to improve their conditions of work.

The increasing power, speed and crowding of machinery were making factories more and more dangerous. Engels, writing of conditions in 1844, remarked that there were so many cripples in Manchester that the inhabitants looked like an army which had just returned from a campaign. It is almost impossible today to realise the indignation with which some of the mill-owners received the suggestion that they should be held responsible for any accidents that occurred on their premises. But no matter how stubbornly they resisted, the tide of opinion was running against them, and, thanks to the combined and persistent efforts of philanthropists, inspectors, statesmen, members of parliament, journalists and others, some effective safety provisions were incorporated in the Factories Act of 1844.

EARLY SAFETY LEGISLATION

The first tangible achievement of the reformers was actually the adoption in 1802 of an Act for the preservation of the health and morals of apprentices and others employed in mills and factories. The inspection of these mills and factories was entrusted to honorary visitors chosen from among the local magistrates and the clergy. An amending Act of 1833, again dealing largely with the textile trades, created a government inspectorate, but it was not until 1844 that clauses relating to the fencing of machinery, the provision of other safeguards and the reporting of accidents were inserted in the Act.

In other countries the lot of the children was sometimes little better. In an account of conditions in the cotton, wool and silk industries in France, compiled in 1840 by the statistician Louis René Villermé, one reads of children of 6 and 8 years of age working in a standing position for 16 to 17 hours a day, badly fed, badly dressed, obliged to walk long distances to the workshop at 5 o'clock in the morning and returning home exhausted at night. In France, too, some zealous reformers, among whom several Alsatian textile manufacturers were conspicuous, strove to mitigate the sufferings of the mill children, and in their efforts originated the movement for the prevention of industrial accidents. Engel Dollfus, who in 1867 founded an association in Mulhouse for the prevention of factory accidents and for the exchange of experience in safety problems, was a man of high social principles, which he expressed in the following words: "The employer owes more than wages to his workers. It is his duty to take care of their moral and physical condition, and this purely moral obligation, which cannot be replaced by any kind of wages, should take precedence over considerations of private interest."

The first piece of factory legislation in France was an Act, dated 22 March 1841, on the employment of children in industrial undertakings, factories and workshops using mechanical power or carrying on continuous processes, and in factories employing more than 20 workers. It also provided for a system of inspection; but safety legislation in the strict sense was not introduced until 1893.

In Prussia the first steps taken to establish a factory inspection system took the form of regulations issued on 9 March 1839, concerning the employment of young workers in factories. A circular of the Prussian Minister of the Interior, Finance and Education, dated 28 May 1845, recommended the appointment of medical inspectors of factories. State factory inspectors, empowered to deal with matters affecting the safety as well as the health of young persons, were appointed for the industrial centres of Düsseldorf, Aachen and Arnsberg in 1853. General protection of workers against industrial accidents and diseases was provided for under the industrial code *(Gewerbeordnung)* of the North German Federation issued in May 1869. A system of inspection covering industrial health and safety generally was introduced in 1872 in Prussia, and at about the same time in the industrial states of Saxony and Baden. An Imperial Act of 15 July 1878 made factory inspection compulsory in all the German states. The industrial accident insurance legislation, under which the system of mutual accident insurance associations was developed, dates from 1884.

In Belgium, industrial safety and health legislation had a rather different origin; it derived from the legislation of the Napoleonic era, partly from inspection legislation and partly from legislation protecting the public from industrial dangers and nuisances. An Act concerning mines, smelting works and similar undertakings, promulgated on 21 April 1810, created a system of inspection and, although the inspectors had no statutory duties in connection with safety and health, they did attend to these matters in practice. Subsequently, pursuant to an Imperial Decree dated 15 October 1810, the Government issued regulations for the protection of the public from nuisances arising from dangerous, unhealthy or obnoxious undertakings, and used them to promote industrial safety and health by treating the workers as members of the public.

Other European countries, including Denmark and Switzerland, had factory legislation on the statute books by 1840, but it was not until much later — in Denmark in 1873, and in Switzerland (at the federal level) after 1877 — that effective systems of factory inspection to enforce safety and health standards were established.

In the United States, Massachusetts was the first state to pass an Act for the prevention of accidents in factories. This Act, dated 11 May 1877, provided for the guarding of belting, shafting and gearing, prohibited the cleaning of machinery in motion, and required elevators and hoistways to be protected and sufficient exits to be provided in case of fire. Massachusetts was also the first state to pass an Act requiring accidents to be reported; it was dated 1 June 1886. Similar Acts were passed by Ohio in 1888, Missouri in 1891, and Rhode Island in 1896.

In the United States, as in Europe, the first factory legislation made no provision for the establishment of special enforcement agencies, on the assumption that complaints would be made by the injured employees. It was, however, found that employees would not make complaints for fear of losing their jobs, and in the 1860s a beginning was made with the appointment of factory inspectors who could conduct prosecutions without calling upon employees to testify. Once again Massachusetts was the first in the field: there a state inspectorate was established in 1867. Wisconsin passed factory inspection legislation in 1885 and New York in 1886. After 1885, too, the principle of employers' liability in respect of employment injury began to appear in the legislation of the different states.

The growing importance and complexity of industry in Western countries, where labour inspection services were responsible for the enforcement of safety laws, made it necessary to add to the staff of these services a number of specialists suitably qualified to cope with the new and increasing number of complicated safety problems. With the assistance of medical, electrical, chemical and other specialists, the labour inspector could become a technical consultant to whom employers and workers could turn, and in this capacity he could make a better contribution to safety promotion than when he was merely an official responsible for enforcing the law.

In some countries, assistance in the promotion of safe working conditions has come from social insurance institutions. These institutions have to pay compensation in case of accidents and are interested in accident prevention as a means of limiting the cost of social insurance. As was stated earlier, workers' compensation is a real cost in terms of the cost of accidents. There can be no doubt that increases in workers' compensation have increased pressure by social insurance institutions to promote accident prevention measures. Their activities in this field have included the issue of enforceable safety rules and the publication of safety pamphlets for different branches of industry. Such a system was followed in Germany after 1884; it resulted in that country having two different state services (labour inspection and social insurance), both of them responsible to some extent for accident prevention — an arrangement that has given rise to some administrative problems.

In the United States, as the number of states to pass laws concerning employers' liability in respect of employment injury increased, the employers' liability was gradually taken over by insurance companies. The latter appointed inspectors to supervise safety measures in the insured undertakings, and in this way they entered the field of accident prevention.

EXCHANGES OF EXPERIENCE ON SAFETY MATTERS

The idea of Engel Dollfus of exchanging experience on safety matters between different undertakings gave a powerful stimulus to the introduction of suitable precautions in industry. Earlier, individual undertakings had sometimes taken far-reaching safety measures, but these had rarely been applied in other factories.

The activities of Engel Dollfus resulted in the adoption of safety measures applicable in all the textile factories in Mulhouse.

In 1889, the Accident Prevention Association in Mulhouse published an album containing all the safeguards known at that time to have given satisfaction in the factories where they were in use – another idea of Dollfus's. It was sent to the international exhibition in Paris, where it received much attention, for by that time safety was considered an important industrial problem in many countries. An improved and expanded second edition was published in 1895. It is most interesting to observe that several of the safety devices described in this manual are still recommended in safety publications.

At about this time, a number of international safety congresses were held (in Paris in 1889, in Berne in 1891 and in Milan in 1894), which had a considerable influence on the legislation of the period. The system of promoting safety by exchange of experience and the publicising of suitable safeguards had been shown to be invaluable but not sufficient to arrive at substantial results. It was found, for example, that those who were responsible for the enforcement of safety measures in the factory itself were not sufficiently independent or had other work to do which prevented them from attending satisfactorily to safety matters; alternatively, there was no co-operation between management and workers, or the workers themselves were opposed to the new measures.

At the Berne Congress, the employers' representatives suggested that these difficulties should be overcome by promulgating safety laws and by setting up state inspection services for their enforcement; in this way safeguarding dangerous places would become a statutory obligation. Moreover, to make sure that the obligation should be a real one, state inspectors, not influenced by local conditions, should be appointed to see that the law was enforced. Three years later, at the Milan Congress, this proposal was put forward again; in addition, it was proposed that governments should promote the founding and functioning of safety associations, organised by private persons, to promote action aimed at improving the safety and protecting the health of the workers. State labour inspectors were to co-operate with these associations.

SAFETY ASSOCIATIONS

As well as the official inspectorates, a number of voluntary organisations have grown up with the objective of improving safety at work.

Most of the voluntary safety organisations in existence are of more recent origin than safety legislation. As far as is known, the oldest safety organisation in the world is the Mulhouse Accident Prevention Association, founded in 1867, to which reference has already been made. Other European countries gradually followed the French example: the Belgian Manufacturers' Association for the Prevention of Industrial Accidents was founded in 1890 and its Italian counterpart in 1894. The Swedish Workers' Protection Association dates from 1905. The British National Safety First Association (known since 1941 as the Royal Society for the Prevention of Accidents) did not come into existence until during the First World War.

The early aim of RoSPA, as it is more commonly known, was to reduce road traffic accidents. However, the aims of the Society have now broadened to include home safety, industrial safety and agricultural safety as well as the original road safety. Another organisation, formed in 1957, was the British Safety Council which promoted education and training among factory employers and workers along similar lines to those of the National Safety Council in the United States. The National Safety Council (NSC), originally the National Council for Industrial Safety, was founded in 1913. At present the NSC is the leading safety organisation in the United States and it is concerned with traffic, home, agricultural, school and industrial safety. Other safety associations include the Cuban National Safety Council, apparently the first of its kind in Latin America, which dates back to 1936. In Asia, the Japanese Industrial Welfare Society (founded in 1928) was the first to appear; the second was the Safety First Association of India, founded in 1931. The National Safety Council of Australia and the New South Wales Safety First Association both began activity in 1927.

TESTING AND RESEARCH INSTITUTIONS

Other types of institutions have grown up with the progress of technology. These institutions are involved in the testing and research of industrial materials and equipment. Mining is the industry for which perhaps the most safety research has been done; gas explosions, dust explosions, fires, electrical equipment and haulage equipment are among the subjects to which an enormous amount of research has been devoted; but industry generally has benefited from research into chemicals, constructional materials, traction equipment, respirators, and many other things.

Some early examples of testing institutions are the Association of Belgian Manufacturers, the German State Material-Testing Institute, the Study and Research Centre of the French nationalised coal industry, the Silicosis Research Institute at Bochum (Federal Republic of Germany), the Italian National Institute for Accident Prevention, the Safety in Mines Research Establishment at Sheffield (United Kingdom) and the United States Bureau of Mines.

QUESTIONS

1. How did the Industrial Revolution give birth to the accident prevention movement?

2. Describe some of the earliest measures taken to prevent accidents.

3. In which parts of the world did accident prevention activities develop first?

4. State what you know of the development of accident prevention legislation and activity in your own country.

ACCIDENT INVESTIGATIONS AND STATISTICS

3

Statistics of accidents have proved to be essential for planning accident prevention measures and for assessing their effectiveness. We have already seen that it is from statistics that we learn how many accidents occur, what kinds of accidents they are, how serious they are, what classes of workers incur them, what machines and other equipment are involved in them, what sort of behaviour is associated with them, and at what times and places they occur most frequently. Statistics provide an overall picture of the situation, and without them it would be practically impossible to estimate needs or judge results.

In order that accurate statistics may be compiled, it is, of course, necessary that all accidents be reported to the person, authority or institution responsible for compiling such statistics. Such reports must provide the kind of information needed for the particular statistical studies in view and in a form that lends itself to statistical treatment. The simplest information refers only to the total number of accidents. If frequency rates are to be compiled, the number of accidents must be studied in relation to the number of hours of exposure to the risk. For the compilation of severity rates, the amount of time lost will also be required. For statistics showing the distribution of accidents by cause, type of accident, nature of injury, equipment involved, or age and sex of the victim, still more information is required, and the more complicated the statistics, the more complicated the report form required. It will often not be possible to fill in a report form until the accident has been thoroughly investigated — which will have to be done in any case if the causes of the accident are to be correctly indicated.

CAUSES OF ACCIDENTS

Before any suitable precautions against accidents can be taken, it is necessary to know exactly how and why they occur. This knowledge has to be obtained by careful investigation of each case. Every accident, even the most trivial, must be investigated.

In countries where social insurance schemes exist or where, for other reasons, accidents have to be reported, accident causes are often defined in such terms as "hand tools" or "falls of objects". These classifications are of little use for accident prevention purposes. More detailed information is needed, and, as a rule, this has to come from a special investigation. Such investigations usually bring to light a series of circumstances or factors, the combination or sequence of which made the accident possible.

Let us look at an example which makes this clear. Suppose a man climbing down a ladder falls because it has a missing rung. Investigation of the accident may reveal the following:

(a) there was a ladder with a rung missing in the workroom;

(b) a worker took that ladder and used it for a small repair job; and

(c) after finishing the job, he came down the ladder without remembering that there was a rung missing.

Each of these three factors was required to bring about a situation in which the accident could happen, but the accident only took place when all three were present in combination. If one of the circumstances could be eliminated, the accident could not recur. When deciding which factor should be considered as the cause of the accident, it is essential that the one chosen is one which can actually be prevented from recurring; this is the only way to achieve practical results in accident prevention.

If the third factor (insufficient attention of the worker) is taken in the first instance, it will be very difficult, if not impossible, to make sure that workers think of their work continuously and never allow their attention to wander, even for a moment. Consequently, this factor should not be considered as the cause of the accident.

The second factor (using a defective ladder) could perhaps be dealt with by giving instructions forbidding the use of defective ladders. Such instructions, however, will not be fully effective, for it will not always be possible to prevent a worker who wants a ladder just for a moment from taking the nearest one to hand instead of spending time looking for a suitable one.

The factor mentioned first (presence of a defective ladder in the workroom) remains to be considered. The accident could easily have been avoided if the management had given orders that every defective ladder should immediately be sent to the repair shop and had seen to it that those orders were carried out. This is the point at which the chain of circumstances could most easily have been broken, and this is the factor that should be considered as the primary cause of the accident. In short, the primary cause may be defined as the most easily preventable circumstance in the absence of which the accident could not have occurred.

Whatever the accident, the identification and evaluation of each component of an accident is a prerequisite for the appropriate preventive measures to be taken.

INVESTIGATION OF ACCIDENTS

The purpose of an accident investigation is to find the causes of the accident in order that appropriate preventive measures can be taken. The National Safety Council in the United States gives the reasons for accident investigation as:

(a) to learn accident causes so that similar accidents can be prevented by mechanical improvements, better supervision or employee training;

(b) to determine the "change" or deviation that produced an "error" that in turn resulted in an accident (systems safety analysis);

(c) to publicise the particular hazard among employees and their supervisors, and to direct attention to accident prevention in general;

(d) to determine facts bearing on legal liability. (An investigation undertaken solely for this last purpose, though, will seldom give enough information for accident prevention purposes. On the other hand, an investigation for preventive purposes may disclose facts which are important in determining liability.)

Basically, whatever type of investigation is undertaken, it must answer the following questions:

Who was injured?

What happened and what were the contributing factors?

When did the accident occur?

Where did the accident occur?

Why did the accident occur?

And finally, and most importantly:

How can a similar accident be prevented from happening again?

There are several methods of conducting an accident investigation which are neither too complicated nor too time-consuming. For minor accidents, good results have been obtained by the following method. The victim goes to the first-aid room and, after treatment, is given an accident investigation form to take to the supervisor; the latter fills it in and sends it to the safety engineer who, according to circumstances, may decide to make a more detailed investigation (or to take some other action) or simply to file it for statistical purposes or for discussion in the safety committee. This method has the advantage of stressing the responsibility of the supervisor for safety in that particular department. However, in many cases, accident investigation forms may be filled in by the victim upon arrival at the medical centre for treatment.

An accident investigation should always be made on the spot. It will be much easier if the investigator finds the situation at the scene of the accident exactly as it was when the accident took place. Consequently, after an accident, the site should be left undisturbed unless changes have to be made to ensure the safety of persons or to prevent further damage.

Whether the site has been disturbed or not, it is desirable to try to reconstruct the sequence of events before, and during, the accident, possibly with the assistance of the injured person, and with the co-operation of witnesses. The investigator should carefully inspect the site and then examine the witnesses. In many cases the causes of the accident will be discovered in this way, but in some cases — e.g. where breakages of metal parts are involved — it may be necessary to seek technical assistance.

If part of a piece of apparatus, such as a chain or a wire rope, breaks, it is desirable to know the cause of the failure, and, for this reason, the material should be examined and tested to discover whether it was unsuitable, had been maltreated or was just worn out. The necessary information may be provided by a microscopic examination, by tests carried out on a sample of the material, or by chemical analysis.

Example. A chain used in a hoisting apparatus with a maximum safe working load of 500 kg broke when 700 kg was lifted. The chain was overloaded, but this circumstance in itself was not sufficient to explain why it broke. Tensile tests on two links showed that they broke when the load was about 2,500-2,600 kg. Links tested by hammering their narrow end until the long side became the short side did not show any defect. A microscopic examination, with magnification of 200 and 1,000, showed ageing phenomena, i.e. changes in the properties of the material which had occurred with the passage of time. These ageing phenomena resulted in decreased resistance to shock, such as that which occurs in working conditions, and this had caused the accident.

When an accident is due to unforeseen chemical reactions, laboratory tests are also necessary in order to ascertain what exactly happened.

Example. A series of inexplicable fires and explosions occurred in a number of dextrin factories. An investigation was made into the chemical properties of the substances present in these factories. First, a general study was made of the influence of small quantities of different substances added to the dextrin on the explosibility of dextrin dust clouds. Next, the circumstances which influenced the composition of dust clouds and the possibility of a dust explosion were studied. These experiments showed that if the air contained some hundredths of a gram of dextrin dust per litre, explosions could be expected; different substances added to the air increased the danger.

It remained to determine the source of ignition. Further experiments were made, which showed that oxidation started at relatively low temperatures (170° C, sometimes even 155° C), and that sufficient heat was developed to raise the temperature to such an extent that spontaneous combustion became possible if heat losses were prevented and sufficient time allowed.

This combination of unfavourable circumstances seldom existed in the factories concerned, but was not unknown, and the research work explained not only why the fires and explosions occurred but also why fires often started on Sundays, more than 24 hours after the closing of the factories on Saturdays.

Although these examples can be used to illustrate the relationship between unsafe acts and unsafe conditions, it is not always so easy to identify human factors in the chain of events leading up to an accident. Human factors, such as the mental and physical state of the employee, may require specialist assistance.

Often accident investigation is concerned with both responsibility and prevention, and this may seriously hamper discovery of the cause. If the persons questioned feel that someone will be blamed as a result of the investigation, those whose consciences are not quite at ease may give incorrect or incomplete information. It may then be impossible to find the cause and consequently to devise means of prevention. In accident investigations, it should always be borne in mind that prevention of accidents is much more important than the mere apportioning of blame.

ANALYSIS AND CLASSIFICATION OF ACCIDENTS

The very many types of accidents which occur make it difficult to develop a method of classification and recording that gives information essential for prevention without being too complicated.

The ILO classification of industrial accidents

In 1962, the Tenth International Conference of Labour Statisticians, convened by the ILO, recommended that, for the study of circumstances surrounding industrial accidents, these accidents should be classified as follows:

A. *Classification of industrial accidents according to type of accident*
1. Falls of persons.
2. Struck by falling objects.
3. Stepping on, striking against or struck by objects excluding falling objects.
4. Caught in or between objects.
5. Over-exertion or strenuous movements.
6. Exposure to or contact with extreme temperatures.
7. Exposure to or contact with electric current.
8. Exposure to or contact with harmful substances or radiations.
9. Other types of accident, not elsewhere classified, including accidents not classified for lack of sufficient data.

B. *Classification of industrial accidents according to agency*
1. Machines:
> 11. Prime movers, except electric motors.
> 12. Transmission machinery.
> 13. Metalworking machines.
> 14. Wood and assimilated machines.
> 15. Agricultural machines.
> 16. Mining machinery.
> 19. Other machines, not elsewhere classified.
2. Means of transport and lifting equipment:
> 21. Lifting machines and appliances.
> 22. Means of rail transport.
> 23. Other wheeled means of transport, excluding rail transport.
> 24. Means of air transport.
> 25. Means of water transport.
> 26. Other means of transport.
3. Other equipment:
> 31. Pressure vessels.
> 32. Furnaces, ovens, kilns.
> 33. Refrigerating plants.
> 34. Electrical installations, including electric motors, but excluding electric hand tools.
> 35. Electric hand tools.
> 36. Tools, implements and appliances, except electric hand tools.
> 37. Ladders, mobile ramps.
> 38. Scaffolding.
> 39. Other equipment, not elsewhere classified.
4. Materials, substances and radiations:
> 41. Explosives.
> 42. Dusts, gases, liquids and chemicals, excluding explosives.
> 43. Flying fragments.
> 44. Radiations.
> 49. Other materials and substances, not elsewhere classified.
5. Working environment:
> 51. Outdoor.
> 52. Indoor.
> 53. Underground.
6. Other agencies, not elsewhere classified:
> 61. Animals.
> 69. Other agencies, not elsewhere classified.
7. Agencies not classified for lack of sufficient data.

C. *Classification of industrial accidents according to the nature of the injury*
10.   Fractures.
20.   Dislocations.
25.   Sprains and strains.

30. Concussions and other internal injuries.
40. Amputations and enucleations.
41. Other wounds.
50. Superficial injuries.
55. Contusions and crushings.
60. Burns.
70. Acute poisonings.
80. Effects of weather, exposure and related conditions.
81. Asphyxia.
82. Effects of electric currents.
83. Effects of radiations.
90. Multiple injuries of different nature.
99. Other and unspecified injuries.

D. *Classification of industrial accidents according to the bodily location of the injury*
1. Head.
2. Neck.
3. Trunk.
4. Upper limb.
5. Lower limb.
6. Multiple locations.
7. General injuries.
9. Unspecified location of injury.

This multiple classification system takes into account that an accident is rarely due to a single factor, but is generally the result of a concurrence of factors. The "type of accident" classification identifies the event which directly resulted in the injury; it indicates how the object, or substance, causing the injury entered into contact with the injured person and it is often regarded as the key to analysis problems. The classification according to agency may be used for classifying either the agency related to the injury or the agency related to the accident. Obviously, more instructive information can be obtained if industrial accidents are classified according to both concepts; however, for accident prevention purposes, the classification according to the agency causing the accident is more important. The classifications according to the nature and bodily location of the injury are designed to provide the necessary information for a detailed analysis.

When causes are being determined, the criterion usually adopted is that of prevention, i.e. the accident is ascribed to the cause which can be eliminated most easily and directly. As we have seen, there are many possible causes of accidents, and some of them, such as psychological factors, are not easy to analyse statistically. Moreover, most accidents are due to a combination of factors—material, physiological, psychological, organisational, educational and other. Consequently, if the greatest possible benefit is to be derived from accident statistics, they must be comprehensive. The classification of industrial accidents according to personal factors is unfortunately very difficult; the difficulty lies in selecting objectively, from among all the personal factors, those elements to which the accident may be attributed. There are only a small number of countries which have built up national classifications of this type.

STATISTICS CONCERNING THE "HUMAN FACTOR" IN ACCIDENT CAUSATION

Statistics have also been compiled to give an idea of how accidents are distributed over the different hours of a working day and how many accidents happen on each day of the week. Such information is very interesting, for it is often assumed that the general environment remains constant and that it is the "human factor" which is much more likely to be the cause of variations. In a British study entitled *2,000 accidents: A shop floor study of their causes* (National Institute of Industrial Psychology, 1971), 2,367 accidents at work were analysed. They occurred in four different types of factories over a period of one and two years. As a rule, it was found that more accidents occurred in the morning than in the afternoon, with a peak time for accidents occurring after mid-morning. This was also the finding of Zetterman in his study of conditions in Sweden in the early 1950s. Figure 5 gives an example of such statistics.

Figure 5.   Percentage distribution of accidents by time of day

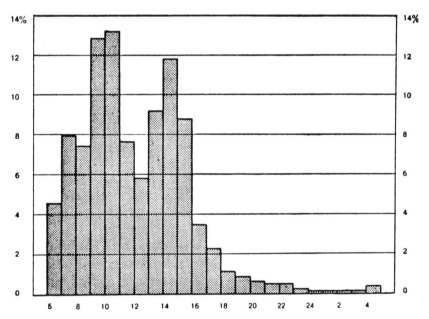

In the study based in the United Kingdom, it was found that local accident peaks occurred before breaks. Although this could have been due to fatigue, it could also have been the result of workers' speeding up production at these times in an effort to meet a target before the break. Furthermore, the accident peak at the end of the afternoon, when people stopped work to tidy up at the end of the day, was either absent or very small.

As to the question of accident distribution over the days of the week, it appears that the highest accident rates are normally found on Mondays and the lowest on Thursdays and Fridays. One underlying factor in this distribution is the absence of workers. In many industrialised countries, absenteeism is always higher on Mondays than on other days of the week. This results in workers having to stand in for absent colleagues and having to undertake unfamiliar jobs that day.

The question of whether more experienced workers have more or fewer accidents than less experienced ones can be discussed with the aid of statistics which indicate how accidents are distributed among workers with different lengths of service, or statistics which give information on accidents in which skilled and unskilled workers, working under similar circumstances, are involved.

Statistics showing the relation between the number of accidents and the age of the workers illustrate another interesting aspect of the influence of the "human factor". Figure 6 gives an example.

Figure 6. Percentage distribution of accidents by age

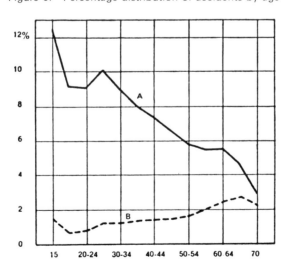

A. Temporary disablement cases
B. Fatal and permanent disablement cases

Recent figures from the United States have revealed that younger workers have more accidents than older workers, and that young male workers have about twice as many accidents as young female workers. For example, one set of figures indicated that workers aged 18-22 made up 7.35 per cent of the workforce but suffered 10.62 per cent of the total number of accidents. Such workers, as well as being young, are new and have little experience of the job.

Statistics of this kind give interesting information on a number of different factors. However, it is difficult to interpret this information accurately, since it is not immediately apparent whether, for instance, the differences shown can be attributed solely to the factors mentioned (age, experience, time of the day, day of the week) or whether other factors are also involved. The difficulty would be partially overcome if the statistics covered sufficiently large numbers of persons, but there would still remain room for doubt on how to interpret them. A certain number of reasonably definite conclusions can, however, be drawn from the statistics shown. Figure 6, for instance, clearly shows the necessity for giving adequate training to young workers. This training must cover the particular job in question.

There can be no doubt that the age of the worker and experience are key factors in accident causation, but it must be remembered that age does not automatically equal experience. Recent studies in the United States have indicated that lack of experience is the most important factor in accident causation.

ARE ACCIDENTS DUE MAINLY TO UNSAFE CONDITIONS OR TO UNSAFE ACTS?

In safety publications, two groups of accidents are often distinguished: those due to technological, mechanical or physical causes, i.e. unsafe conditions, and those due to unsafe acts. To the first group belong accidents caused, for example, by defective parts, unguarded machines, damaged electric cables and worn-out hoisting ropes. To the second group belong those resulting from absent-mindedness, negligence, foolhardiness, or ignorance of risk. The first group is often considered to comprise 15 per cent of all accidents, the second 85 per cent, and the conclusion is accordingly drawn that attention should be concentrated on the latter group.

In one of the earliest reports on the subject, Heinrich, in his 1928 study of 75,000 accidents, established the much-quoted 88 : 10 : 2 ratio. This meant that 88 per cent of all accidents were caused by unsafe acts, 10 per cent by unsafe conditions and 2 per cent by conditions which could not have been prevented. However, in this classical study, only one cause was reported for each accident, and, as we have already seen, accidents are the result of a number of interdependent factors. Also, when Heinrich recognised the accident as a result of unsafe acts and unsafe conditions, he selected what he considered to be the major cause. Consequently, this ratio is now considered by many to be invalid. If anything, the ratio is far more likely to be nearer 1 : 1 between unsafe acts and unsafe conditions.

Let us examine this in more detail. Many accidents which at first sight are placed in the larger group might, on further examination, equally well be placed in the smaller group. For example, a worker climbs to the very top of a ladder, stretches up to do his task, loses balance and falls. At first sight, this is an unsafe act, and clearly it is. However, with a longer ladder, the worker would not have had to stretch up. Thus, this is an unsafe condition.

An accident is very seldom due solely to unsafe behaviour. As already stated, accidents are usually caused by a group of circumstances; one of these may be unsafe acts, but probably unsafe conditions are present as well, and so it would be equally justifiable to classify the accident as due to unsafe mechanical or physical conditions.

In practice, it will be possible to classify the great majority of accidents in such a way that an unsafe act by a worker is not given as the primary cause (i.e. the factor on which efforts to prevent a recurrence of the accident should be concentrated). The following example is an illustration of how this is done.

Example. A 15-year-old boy had the job of cleaning the gangways of a workroom and was told not to clean under the machines. When he saw oil on the floor under a rope-making machine, he cleaned that part of the floor also, but, as he did so, the cotton waste used for cleaning became caught between two gear wheels just above the floor and protected by a hood on the top and sides. As he tried to pull out the cotton his hand was caught between the gears and badly mutilated.

For prevention purposes, it is sufficient to concentrate on the fact that the rope-making machine was inadequately guarded and to make it a general safety rule that gears shall be completely enclosed. It will be more difficult to correct factors relating to wilful disregard of instructions and to operations conducted without authority.

It has already been pointed out that no useful purpose will be served by citing circumstances which cannot be prevented; hence carelessness, negligence, absent-mindedness and the like should not be considered as main accident causes, although they may, as already pointed out, be contributory factors. There are so many different circumstances which disturb the mind of a worker but cannot be eliminated—a quarrel with a fellow worker, a difference of opinion over wages with the supervisor, poor health, difficulties at home, day-dreaming—that it is impossible to prevent moments of carelessness or absent-mindedness. We shall return to this point in the eighth lesson when discussing the psychological aspects of accident prevention.

SERIOUS ACCIDENTS, MINOR ACCIDENTS AND NEAR ACCIDENTS

Earlier it was stated that every accident, however trivial at first sight, should be investigated. For many years, only accidents causing serious injuries have been investigated and minor accidents have largely been ignored.

Early statistics on the subject show that one accident involving major injury happens for every 29 accidents resulting in minor injuries and for every 300 accidents which do not cause injury (i.e. "near accidents").

Some investigators give the ratio as 1 : 20 : 200. Others quote much lower figures depending on the type of industry. In heavy industry, where there is theoretically more likelihood of sustaining a major injury, the ratio will be low. The opposite is true in light industry where the ratio of minor to major injuries may be 50 to 100 times higher.

Whatever the figures, the point is that, for every major accident, many dangerous incidents occur which do not cause injuries. This knowledge can be used to great advantage in planning safety programmes; for if sufficient attention is paid to accidents in which no injuries occurred, there is every likelihood that the number of accidents resulting in injuries, and especially serious injuries, will fall. In any case, it will be necessary to pay special attention to minor accidents and near accidents, because often the seriousness of an accident is not in any way an indication of the frequency with which it will happen again; nor does the fact that an accident did not cause any injury on one occasion constitute an assurance that, under similar circumstances, a serious accident will not occur in future. Consequently, it would be very wrong if, out of the 330 accidents mentioned at the beginning, measures should be taken to prevent the recurrence of only one of these accidents and the other 329 should be ignored. The most important task is that of finding ways of discovering and preventing the 300 near accidents.

It is easy to arrange for lost-time accidents to be reported to the safety engineer or any other appropriate official, and most of the 29 minor injuries can be reported by the first-aid department; but what is to be done about the 300 near accidents? Some undertakings consider it important that these should also be reported, for among them are cases of stumbling, slipping or falling which might have resulted in an injury, even a serious injury, but which just by chance did not. If these events could be discovered, it might be possible to take steps to prevent their recurrence, and the number of accidents in the minor injury and lost-time groups could probably be reduced.

In undertakings where the importance of accidents of this type is realised, attempts have been made to solve the problem of how to learn from them and not apportion blame. In one undertaking, two or three selected workers in each department are responsible for reporting all small defects and shortcomings in their area, such as a hole in the floor, a split-pin replaced by a nail, or a broken window cord, to the safety engineer or the safety committee. The defects are then repaired by the maintenance department. In another undertaking, the safety engineer must obtain information on no-injury accidents at meetings with the supervisors.

COMPILATION OF ACCIDENT STATISTICS

Statistics may be compiled for a single undertaking, a region, an industry or all the industries in a country. Specialised statistics may be compiled for particular types of accidents (e.g. electrical accidents or ladder accidents), for particular classes of workers (e.g. young persons) or for other types of information. As we have already seen, statistics of the same kind for different years serve to show whether the number of accidents is increasing or decreasing, and hence how successful or unsuccessful accident prevention work has been in the undertaking, region or industry concerned. Statistics prepared for different undertakings working under more or less the same conditions indicate whether a certain undertaking is better than average or whether it needs substantial improvement from the accident prevention point of view.

It is clear, therefore, that accident statistics should be comparable not only from year to year but also from industry to industry, region to region and, if possible, country to country. The principal limitation on the comparability of accident statistics lies in the dual purpose for which they are designed: use in accident prevention, and use in accident compensation. For prevention purposes, statistics of accidents should provide complete information on cause, frequency, industry and occupation, as well as on other factors that influence risk. Statistics for compensation, on the other hand, are used mainly for administrative purposes and must show the number of accidents of each degree of severity, the length of disability and the amounts paid in compensation; for these purposes, various legal conditions to which the granting of compensation is subject are incorporated in the definition of an accident. Failure to appreciate the distinction between these two uses has, in the past, proved a serious obstacle to the use of accident statistics for purposes of prevention. Statistics to be used for accident prevention should not be designed mainly to meet the requirements of workers' compensation authorities.

If accident statistics are to have the highest possible degree of comparability for prevention purposes, the following principles must be applied:

1. Accident statistics should be compiled on the basis of a uniform definition of industrial accidents, framed for the purposes of prevention in general, and the measurement of the importance of the

risk rates in particular. All accidents as thus defined should be reported and tabulated uniformly.

2. Frequency and severity rates should be compiled on the basis of uniform methods; there should be uniform definitions of accidents, uniform methods of estimating the time of exposure to risk, and uniform methods of statement of the risk rates.

3. The classification of industries and occupations for the purposes of accident statistics should be uniform.

4. The classification of accidents according to the circumstances in which they occur, and according to the nature and location of the injuries, should be uniform, and the principles used for determining the criteria for selection should be the same in all cases.

It is not absolutely essential that national statistics should be comparable in every detail, but they should be comparable on essential points; in each country, data in addition to those called for in international comparisons can be compiled for specific purposes.

Some progress has been made towards the international standardisation of industrial accident statistics, and measures have been planned, particularly under the auspices of the International Labour Office, which has been interested in the problem ever since the First International Conference of Labour Statisticians in 1923. Various International Conferences of Labour Statisticians have examined the problems of industrial injury statistics. The Tenth Conference, convened by the ILO in 1962, adopted a resolution establishing basic international standards in the field of employment injuries. In addition to giving international definitions, for statistical purposes, of fatalities, permanent disablement and temporary disablement, this resolution embodies the four classifications mentioned above and includes recommendations on the calculation of frequency and incidence rates and the classification by industry, duration of incapacity for work and other characteristics, such as sex, age, occupation, skill and experience, the day of the month and the month of the year, the time of the accident in respect of the work schedule, the size of the establishment, and so on.

CALCULATION OF ACCIDENT RATES

To compare the number of accidents in one factory with that in another in the same branch of industry, it is necessary to take into account the differences which may result from the differences in the numbers of workers employed in the two factories. This can be done by calculating the accident frequency rate, i.e. the number of injuries for each million work-hours of exposure. This is expressed by the following formula, in which *F* represents the frequency rate:

$$F = \frac{\text{number of injuries} \times 1,000,000}{\text{total work-hours of exposure}}$$

Example. An undertaking with 500 workers, working 50 weeks of 48 hours each, had 60 accidents causing injury during one year. Owing to illnesses, accidents and other reasons, the workers were absent during 5 per cent of the aggregate working time. Thus, the total number of work-hours (500 x 50 x 48 = 1,200,000) has to be reduced by 5 per cent (60,000), giving the real number of work-hours of exposure as 1,140,000. This being so —

$$F = \frac{60 \times 1,000,000}{1,140,000} = 52.63$$

This frequency rate indicates that, in one year, about 53 accidents causing injury occurred per million work-hours.

So far, only the number of accidents has been considered, and this is not a very exact measure of the effects of accidents. To obtain a better idea of the situation, the severity rate must also be calculated. The international resolution on methods of compilation of severity rates, adopted in 1947 by the Sixth International Conference of Labour Statisticians, was not retained by the Tenth Conference in 1962. This was mainly because some countries calculate the severity rate on the basis of the total number of days lost per thousand work-hours of exposure, other countries use the time expressed in days per million work-hours of exposure, and yet other countries use as a denominator the average number of employees or insured persons, or 300-day work-years.

Example. If, in the example given for the calculation of the frequency rate, the number of days lost as a result of the 60 accidents was 1,200, the severity rate *(S)* would be as follows:

$$S = \frac{1,200 \times 1,000}{1,140,000} = 1.053$$

This means that in a year about one day was lost per thousand work-hours of exposure, or 1,053 days per million work-hours of exposure, or, on the basis of 2,400 hours of work per year, 2.4 days per worker.

The calculation of severity rates is more difficult when an accident gives rise to permanent disability or death. To cover such cases, there is usually a national schedule specifying the number of days to be counted as lost (time charges) for statistical purposes for each type of disability. However, there are considerable differences in the scales used by different countries to assess the number of days lost as a result of injuries involving permanent partial disability or permanent total disability and death.

The Tenth International Conference of Labour Statisticians recognised that the lack of a uniform scale of time charges constituted one of the main obstacles to the international comparability of severity rates, and concluded that further research was needed before an international method could be recommended.

Example. If, in addition to the 60 lost-time accidents mentioned in the previous examples, one fatal accident had occurred, the frequency rate would have been —

$$F = \frac{61 \times 1,000,000}{1,140,000} = 53.5$$

The time lost in days in that case, using a national schedule specifying a time change of 7,500 days for fatal cases, would have been 8,700 (7,500 + 1,200) and the severity rate —

$$S = \frac{8,700 \times 1,000}{1,140,000} = 7.63$$

As might be expected, such a serious accident has a considerable effect on the severity rate, but does not greatly affect the frequency rate.

Accident frequency and severity rates give valuable information on the safety situation in a factory, both absolutely and by comparison with other factories working under similar conditions. It is therefore desirable that rates should be published regularly for different branches of industry.

At present, few countries publish severity rates for industrial accidents. It is difficult to make comparisons between the countries that do publish such rates because of the different methods used to calculate the number of days lost in the event of death and permanent disability, whether it be total or partial.

PRESENTATION OF STATISTICS

Accident statistics are not compiled solely for research and study in the interests of accident prevention. Although this is the main reason, it is also important to give all persons concerned appropriate information on the accident situation, in order to warn them of the dangers to which they are exposed, keep their interest alive, and make them safety-minded. Accordingly, it is sometimes desirable to present statistical data not only in figures but also in pictures. The latter often attract far more attention than figures and are the only means of making statistics understandable to persons who cannot read or write. In countries where a large part of the population is illiterate, the publication — possibly by trade unions — of sketches and drawings giving facts about accidents and their effects might well prove an extremely effective instrument for instilling safety-mindedness into the workers. Some examples of accident records in pictorial form are given in figures 7, 8, and 9.

Figure 7. Total number of accidents per year, by cause (in thousands)

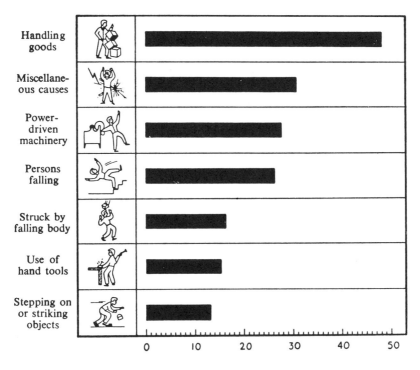

Figure 8. Number of fatal accidents per year, by cause

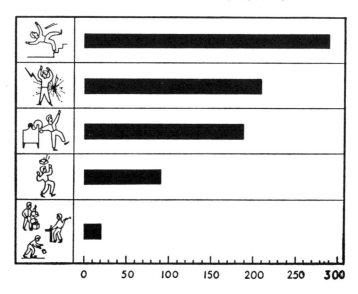

Figure 9.   Percentage distribution of injuries, by part of body affected

15 %
8 %
10 %
3 %
15 %
15 %
34 %

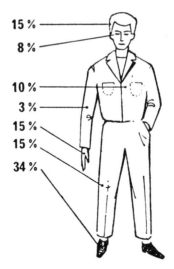

QUESTIONS

1. What kinds of occurrences can be considered to be causes of accidents?

2. What purposes are served by investigating accidents?

3. What do you think of the classification of accident causes recommended by the ILO?

4. Are so-called technical and human causes of accidents related? If so, in what ways?

5. What is the significance of near accidents for accident prevention purposes?

6. Mention some of the steps that must be taken so that accident statistics can be compiled on a uniform basis.

7. What is meant by accident rates?

8. Describe the principal uses of the different kinds of accident statistics.

SOME PRINCIPLES OF ACCIDENT PREVENTION I. FIRE AND EXPLOSION

4

It is well known that death from fire is far more common in the home than in industry. In the United Kingdom, it is estimated that there are between 600 and 800 deaths per year as a result of fire, and, of these, less than 5 per cent are due to industrial fires. In strictly monetary terms, these industrial fires cost about £2 million of damage to property for every person killed by the fire.

Many factory fires and explosions occur outside normal working hours. In such cases there is a reduced risk of personal injury but the resulting loss of employment makes such fires both an economic and a social calamity.

Example. One Saturday, in June 1974, a chemical works was virtually demolished by an explosion of warlike dimensions. Of those working on the site at the time, 28 were killed and 36 others suffered injuries. If the explosion had occurred on an ordinary working day, many more people would have been on the site, and the number of casualties undoubtedly would have been far greater. Outside the works, injuries and damage were widespread but no one was killed. Fifty-three people were recorded as casualties by the casualty bureau which was set up by the police; hundreds more suffered relatively minor injuries which were not recorded. Property damage extended over a wide area, and a preliminary survey showed that 1,821 houses and 167 shops and factories had suffered to a greater or lesser degree.

Fires occurring during working hours constitute a much greater danger to workers.

Example. In 1956, an old four-storeyed factory building containing worsted spinning and twisting machines, in which 41 persons were working, was destroyed by fire; eight persons, all working above the ground floor, were killed and three injured. The fire started on the ground floor, when the flame of a plumber's blowlamp ignited waste wool fibres on the ceiling joists, and spread rapidly from one end to the other. The factory was properly equipped with an outside fire escape, but the access to it was soon cut off by the flames; the remaining internal staircase was steep and narrow, and afforded an easy passage for the dense smoke rising from the ground floor — for, although it was lobbied off from the upper floors, it was open at ground level. Workers escaping down these stairs panicked at the sight of the smoke, ran back up the stairs and lost their lives — not by burning, but by asphyxia.

Much can and should be done to prevent disasters of this kind by those responsible for the building of factories; but the workers, too, have a very real responsibility for ensuring the effectiveness of fire-prevention measures.

Fires and explosions can also affect the community surrounding the factory in both the short and the long term — as, for instance, when an explosion in a chemical works gives rise to long-term pollution effects in the immediate neighbourhood.

COMMON FIRE HAZARDS AND THEIR PREVENTION

For a fire to start, three elements must be present: oxygen (from the atmosphere), combustible material (fuel), and heat (this is essential to ignite the fire, but, once the fire has started, it produces enough heat to maintain itself). If any one of these components is removed, the fire will not start, or, if the fire is already in progress, it will go out.

Figure 10.   How fires start

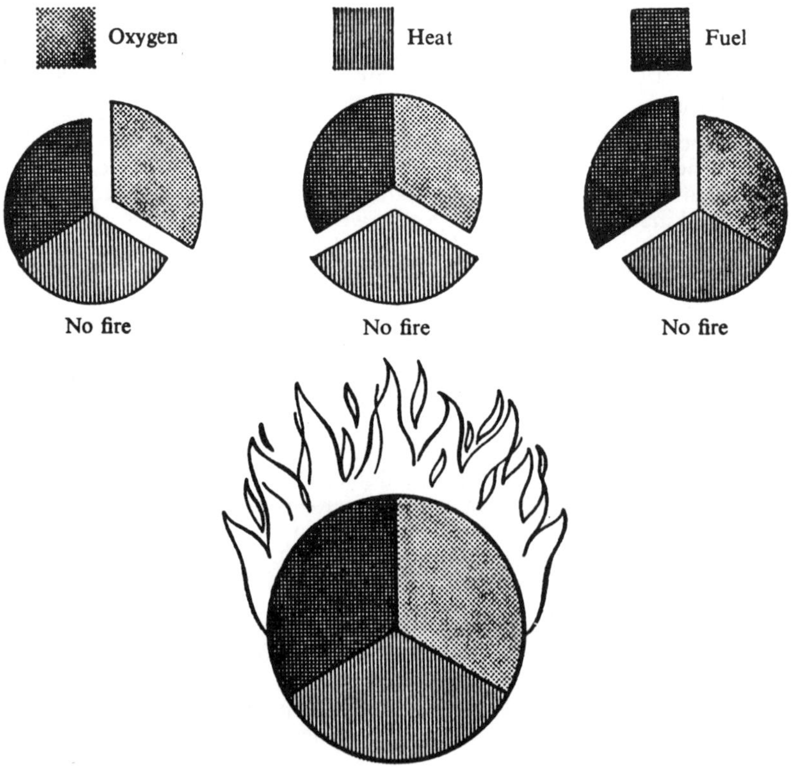

Oxygen   Heat   Fuel

No fire   No fire   No fire

Thus, methods of fire prevention basically involve reducing or eliminating one of these components. In almost every industrial situation, two of the three components are usually present — oxygen and fuel. Therefore, it is essential to ensure that the third component — heat — is never sufficient to start a fire.

Let us look at some of the common causes of fires and at some simple means by which they can be prevented. In a recent analysis of over 25,000 fires reported to the Factory Mutual Engineering Corporation over a ten-year period, the following data were obtained on the ignition sources of fires.

The leading cause of industrial fires was *electrical faults* (23 per cent). Most of these fires started in wiring and motors, and most of them could have been prevented by proper maintenance.

Example. In 1980, a fire and series of explosions occurred at a warehouse. On the morning of the fire, the warehouse contained some 49 tonnes of liquefied petroleum gas in cartridges and aerosol containers, as well as about 1 tonne of petroleum mixtures in small containers, raw materials and packaging materials. It was almost certain that the source of ignition was the electrical system of a battery-operated fork-lift truck.

Smoking caused 18 per cent of industrial fires, but, of course, it is a potential cause of fire everywhere. One very common fire precaution is the "no smoking" rule. In practice, however, it is not always observed, for some persons find it extremely difficult not to smoke during the four or five consecutive hours of a shift. In factories where no special fire risks exist, such as metalworking establishments, the solution sometimes adopted is to prohibit smoking only during the last working hour, in order to ensure that no unextinguished cigarette ends are left lying around unobserved after working hours, when they could start a fire. In factories where the fire risk is particularly great, such as textile plants, a special room is sometimes provided where workers are allowed to go during working hours for an occasional smoke. Thus, prevention methods are largely a matter of education and the provision of specified smoking areas.

Friction accounted for about 10 per cent of industrial fires and was largely the result of badly maintained, and hence overheated, machines.

Overheated materials caused 8 per cent of these fires and usually involved overheating flammable liquids and materials.

The other causes given in this report were: hot surfaces (7 per cent, heat from boilers, furnaces, etc.), burner flames (7 per cent, improper use of portable torches, etc.), combustion sparks (5 per cent, sparks and embers released from incinerators, etc.), spontaneous ignition (4 per cent, poor housekeeping and poor removal of waste), cutting and welding (4 per cent, sparks, arcs and hot metal from cutting and welding operations), exposure (3 per cent, spread of fire from neighbouring properties), incendiarism (3 per cent, fires started on purpose), mechanical sparks

(2 per cent, sparks from foreign material in machines), molten substances (2 per cent), chemical action (1 per cent), static sparks (1 per cent), lightning (1 per cent) and miscellaneous (1 per cent).

Whatever the cause, fire prevention largely involves elimination of these ignition sources because the other two components, fuel and oxygen, are so often present. Elimination may involve using pneumatic equipment rather than electrical equipment or building safety devices into the system.

STRUCTURAL FEATURES AND EXITS

The first line of defence against fire is in the construction of the building itself. Industrial buildings should be sufficiently fire-resistant, with respect to the fire risks inherent in the processes carried on inside. This is, of course, primarily a matter for architects and designers; but the workers themselves can give invaluable assistance on some aspects of the problem (for instance, the fourth and fifth points in the list of recommendations below). Fire-resistant construction should ensure that structural parts do not readily burn and that fires cannot spread, either horizontally or vertically, through walls, floors, doors, elevator shafts, stair wells, or ventilating ducts. Exits are most important, and should conform to the following general rules:

1. No part of the building should be far from an exit leading to the outside, the distance depending on the degree of hazard.
2. Each floor should have at least two exits, sufficiently wide, protected against flames and smoke and well separated from each other.
3. Wooden stairs, spiral stairs, lifts (elevators) and ladders should not count as exits.
4. Exits should be signposted and well lit.
5. Exits should always be kept unobstructed.
6. Outside stairways and fire escapes should not lead into interior courtyards or blind alleys.

FIRE-EXTINGUISHING EQUIPMENT

Fire-extinguishing equipment may range from buckets of water or sand to complete sprinkler systems. The type and amount of equipment needed will depend on the size and construction of the building to be protected and the processes carried on in it. Sometimes it is sufficient to have portable fire extinguishers, or even a supply of dry sand or some barrels filled with water. Most factories in regions with piped water supplies will have hydrants and hoses. To rationalise the choice of fire-extinguishing equipment and the precautions that should be taken, a number of countries recognise various classes of fire.

Class A fires (carbonaceous solids – wood, paper, rubbish, etc.). The usual method of extinguishing this type of fire is by water jets which quench the fire and cool the material to below its ignition temperature. Care must be taken that sufficient time is allowed for the water to penetrate and cool the whole of the material otherwise the fire may start up again.

Class B fires (flammable liquids or liquefiable solids – solvents, oil, paint, etc.). This type of fire is not so easy to deal with because the methods used to put out the fire depend largely on the nature of the flammable liquid. If the liquid is insoluble in, and lighter than, water, the use of water jets will simply spread the fire to the surrounding areas. The liquid will float on the surface of the water, just like burning oil on top of the sea. If the liquid has a low flashpoint, its vapour forms an explosive mixture with air and does not remain localised. Thus, a source of ignition some distance away will cause a sheet of flame extending back to the liquid itself. Normally a smothering or blanketing method (foam, etc.) is used to put out this type of fire.

Class C fires (gases – fractured gas main, etc.). If possible, the best way of dealing with a fire of this type is to try to stop the gas leak, whether it be from a fractured gas main or from a leaking gas cylinder.

Class D fires (metals – magnesium and its alloys, and sodium and potassium if in contact with water). A dry powder is required to extinguish this type of fire and the choice of powder depends on the metal that is burning. One of the greatest problems in these fires is toxic fumes from the metals.

Care should be taken that portable fire extinguishers do not constitute a danger in themselves. This is sometimes the case with appliances of unsuitable construction containing chemicals which may block the spray nozzle. When such an appliance has to be used, a seal is broken or the appliance is turned upside down and the chemicals inside are mixed. The pressure inside the cylinder then increases, forcing out a jet or spray of extinguishing material; but, if the nozzle is blocked, the appliance will burst. Suitable construction and regular checks should prevent these accidents. Another serious risk arises when extinguishers are filled with toxic substances such as methyl bromide or carbon tetrachloride; there will be a risk of poisoning if the extinguisher leaks or if it is used in a confined space. Such extinguishers should therefore not be used indoors.

Fire hoses provided with nozzles should be available where practicable, and it is important that the connections should fit the equipment of public fire brigades so that the latter can operate in the factory.

Factories where the risk of fire is great, such as textile plants, should normally be equipped with sprinkler systems. In such systems, water under pressure is carried in a network of pipes near the ceilings of the workrooms. The pipes have nozzles closed by metal strips. If a fire breaks out, the heat melts the nearest strip and water is sprayed into the room.

FIRE ALARMS

Every workplace should have an alarm system to warn people if a fire starts. The alarm system can be automatic, or alarm bells, whistles or sirens can be installed in different parts of the factory, with push-buttons or handles in all workrooms to operate the alarm if necessary. Alarms must be audible everywhere in the factory, including workrooms, storehouses, gangways, locker rooms, lavatories and washrooms.

FIRE-PREVENTION ORGANISATION

There is more to fire protection than fire-resistant construction and the provision of fire-extinguishing equipment. The workers themselves have an important part to play in the organisation and training of fire brigades, fire drills, and the inspection and maintenance of fire-fighting equipment.

Firstly, every undertaking should have trained fire-fighters on each shift; large undertakings may have complete brigades and, if the risk warrants it, a full-time fire-prevention officer. Fire-fighters should be kept in training by regular drills.

Secondly, it is essential that undertakings be inspected at suitable intervals for fire risks and to see that all fire-fighting equipment is in good condition. Some undertakings employ a full-time watchman for this purpose.

Thirdly, it is desirable that regular fire-drills be organised to make sure that all workers know how to use the fire-extinguishing equipment, where the nearest exit is, and how to leave the building calmly. During fire-drills, fire officers should check that there are enough exits for the factory to be evacuated quickly. However, it should not be overlooked that fire-drills are expensive, as they suddenly stop production and it may take some time for work to be resumed. They should therefore be held judiciously.

Lastly, it is also necessary to ensure close co-operation with the local fire brigade. In some factories, the telephone number of the fire brigade is indicated near every telephone, so that anybody can ask them to come if necessary. Some factories, however, arrange for workers to call the factory telephone exchange, which then calls the fire brigade. The practice of authorising the factory telephone operator to call the fire brigade only if asked to do so by certain specified members of the factory personnel is undesirable; it has caused delays that have sometimes proved to be extremely costly.

PRECAUTIONS AGAINST EXPLOSIONS

In some factories precautions are necessary not only against fire risks but also against the risk of explosions, which are usually very violent and destructive.

Explosions may be caused by commercial explosives or by concentrations of certain vapours, gases or dusts in the air. Trinitrotoluene, fulminate of mercury and lead azide are examples of commercial explosives. Dusts that may be explosive when mixed with air include organic dusts, such as those of flour, grain, sugar, starch and cork, and some metallic dusts, such as those of aluminium and magnesium. Among the vapours and gases that may cause an explosion when mixed with air are acetylene, butylene-n, carbon monoxide, ether, hydrogen sulphide and methanol. Not all mixtures of such gases and vapours with air are explosive; the mixture must contain a certain proportion of the ingredients. This proportion lies between what are known as the upper and lower explosive limits. For instance, any atmosphere containing at least 1.5 per cent and at most 100 per cent of acetylene is explosive.

Example. In a basement, a worker was repairing an ammonia refrigerator. Suddenly, a large quantity of the gas escaped from the apparatus and the mixture of air and ammonia was ignited by an open gas flame. The resulting explosion completely destroyed the ground floor of the building.

For ammonia, the lower explosive limit is 15 per cent by volume and the upper limit is 27 per cent by volume.

Mixtures of petrol vapour and air have caused many explosions in such places as car repair workshops.

Example. During repair work on a car, petrol was spilt on to the floor. Petrol vapour spread through the workroom and into a small office, where it was ignited by an open electric radiator used for heating the office. The whole building was burned down.

Explosive limits for petrol tend to be variable but are in the order of 1.4 per cent by volume for the lower limit and 7.6 per cent by volume for the upper limit.

A dust explosion occurs when a suitable mixture of flammable dust and air is ignited by a heat source of sufficient intensity — for instance, when a cloud of dust is ignited by a flame, a spark or a very hot object, such as a carbon-filament electric bulb. Dangerous concentrations of dust may be present in the air, for example, in workrooms, pneumatic conveyors, milling equipment and dust exhaust systems. The sources of

49

ignition may be open flames, poorly maintained power transmission equipment, unsuitable electrical installations, static electricity, and even smoking.

Example. In a flour mill, large quantities of dust had settled on beams, window-sills and other parts of the building. When a small fire started one day, workers tried to extinguish the fire with a hose. However, the force of the water jet whipped up a cloud of dust which was ignited by the fire and caused a minor explosion. This explosion shook loose all the dust which had settled on beams and other parts of the building. The result was a second explosion, this time sufficiently powerful to destroy the entire plant.

A great number of precautions are necessary in the manufacture, handling, storage and use of commercial explosives. They will not be dealt with here.

It is best to prevent the formation of explosive air-gas and air-vapour mixtures, but, if this is not practicable, they can be diluted beyond the lower explosive limit by general ventilation, or removed at source by local exhaust ventilation.

Where there is a risk of dust explosions inside apparatus, the oxygen content of the air may be reduced by removing some air and replacing it by an inert gas such as nitrogen or carbon dioxide.

Preventing dangerous dust concentrations from forming in dust exhaust systems is primarily a matter of proper design. But correct operation may reduce the risks still further. For instance, it is advisable to keep the exhaust machinery running a few minutes after the machines it serves have stopped. In this way, the ducts will be cleaned, and there will be no danger of explosive clouds of dust being whirled up in them when the exhaust fans are started.

QUESTIONS

1. Mention some common fire hazards and means of eliminating them.

2. What requirements should building exits satisfy?

3. What precautions are required with portable fire extinguishers?

4. How should fire protection be organised in a factory?

5. Name some explosive dusts.

SOME PRINCIPLES OF
ACCIDENT PREVENTION
II. MACHINE GUARDING

5

It is customary to divide machinery into a number of categories, namely prime movers, transmission equipment and working machines, all of which exhibit considerable variety. The particular hazard depends on the type of machine, its function and its mechanical motion. It is not therefore possible to deal with machine guarding in detail here. The guarding of even a single machine may be complicated if it has belts, gears and a number of different tools. This lesson will accordingly be confined to the general aspects of the problem of guards.

In the early days of the Industrial Revolution, it was factory machinery that caused the spectacular accidents which so aroused public opinion; and some of the earliest measures to introduce safety legislation and inspection were designed to reduce the dangers inherent in machinery. Machinery is still important from the accident-prevention point of view: in highly industrialised countries, machine accidents account for more than 30 per cent of all permanent partial disabilities and about 9 per cent of all permanent total disabilities and fatalities.

The practice of fitting guards to machines spread gradually; but the guards were often unsatisfactory for one reason or another — they were unreliable, they hampered the work, or they needed too much attention. Consequently, the guards were often taken off and work went on with unprotected machines.

Usually, the designers of guards were mainly concerned with compliance with the law, or removal of a risk, and gave little thought to the influence of a guard on production, or to the nuisance it might represent for the worker. Sometimes — as in the case of the fencing of dangerous parts of power-transmission equipment — this attitude did not matter much; but in other cases (e.g. with woodworking machinery and metal presses), the guards designed seriously hampered efficient production. The result was that they were often removed, replaced whenever an inspector came by, and removed again as soon as he had left the factory. Work was done on these machines without guards, and the machines remained as dangerous as before. Sometimes, machines would be produced with the same style of guard for as long as 20 or 30 years; but these guards would never be used. Thus, the fact that a particular type of guard remained in use for a long time did not necessarily mean that it had proved its worth.

In some countries, machine guarding has been promoted by the setting up of committees to study means of protecting a particular type of machine. Such committees often consist of representatives of the labour inspectorate, the social insurance authority, machinery manufacturers and purchasers, and workers. In the United Kingdom, for instance, they have produced ideas for the protection of metal presses and of machinery used in the textile and rubber industries. The committee system has been used in the Netherlands to study the protection of lift machinery, the transportation and storage of flammable liquids, and other matters. It has proved valuable, not only for dealing with difficult technical problems, but also when adequate safety precautions have been an important factor in the cost of equipment, as is the case with lifts. In addition, this way of dealing with safety problems does much to ensure the co-operation of all concerned when recommendations have to be put into practice.

In France, the method of official certification is used. The competent authority lays down the general principles to which the protection of a particular type of machine must conform. Manufacturers of safety equipment have to submit their devices to a committee. If the devices are up to standard, the committee certifies that they conform to the general principles governing the protection of the machine concerned. Once a protective device has been certified in this way, it can be sold and used.

The question of sale or transfer of machinery has generated considerable interest over the past few years. The Guarding of Machinery Convention, 1963 (No. 119), adopted by the International Labour Conference at its 47th Session, specifically pointed out that the sale and hire of machinery of which the dangerous parts were without the appropriate guards, should be prohibited by national laws or regulations or prevented by other equally effective measures. Yet even today, few countries, including many of the leading industrial nations, have ratified this Convention, and unguarded machinery is still being produced and sold. Furthermore, it is not uncommon for such machinery to be sold to the developing countries as they become more industrialised.

CONDITIONS TO BE SATISFIED BY GUARDS

The purpose of machine guarding is to prevent any part of a worker's body or clothing from coming into contact with any dangerous moving part of a machine. There are several ways of doing this, depending on the nature of the hazard. Thus, a machine may be so designed and built that all the potential danger zones are enclosed or covered. Dangerous operations may be fully automated so that the worker does not have to come into contact with any hazardous parts. Machine guards, and other protective devices, may be attached to the machine. Whatever method is used, a successful machine guard is one that allows a worker to operate a machine easily, without risk or fear of injury.

For convenience, the requirements that a machine guard should satisfy are analysed here on the basis of the *Model code of safety regulations for industrial establishments*, drawn up by a tripartite technical conference organised by the ILO in Geneva in 1948 (a revised edition is in preparation at the time of writing). The *Model code* was for governments and industries to use as they thought fit in the drawing-up of their safety regulations and rules. Regulation 82 of the *Model code* reads as follows:

1. Guards should be so designed, constructed and used that they will—

(a) provide positive protection;
(b) prevent all access to the danger zone during operations;
(c) cause the operator no discomfort or inconvenience;

Figure 11. Power press with guard providing positive protection. An interlocking mechanism prevents the die from coming down as long as the guard is not closed

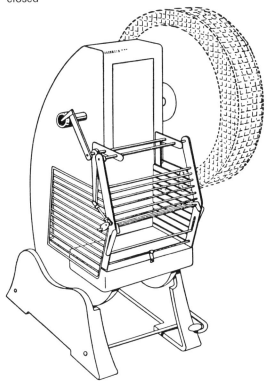

(d) not interfere unnecessarily with production;
(e) operate automatically or with minimum effort;
(f) be suitable for the job and the machine;
(g) preferably constitute a built-in feature;
(h) provide for machine oiling, inspection, adjustment and repair;
(i) withstand long use with minimum maintenance;
(j) resist normal wear and shock;
(k) be durable, fire- and corrosion-resistant;
(l) not constitute a hazard by themselves (without splinters, sharp corners, rough edges, or other sources of accidents); and
(m) protect against unforeseen operational contingencies, not merely against normally expected hazards.

(a) *The guard should provide positive protection.* This means, for example, that, should the guard cease to operate for any reason, the machine will automatically stop, or access to the danger zone will be prevented. As the name implies, positive protection means that the worker is actually prevented from coming into contact with the danger areas of the machine. Such an arrangement is illustrated by figure 11.

(b) *The guard should prevent all access to the danger zone during operations.* It is not sufficient for the guard to give a warning signal, for instance by means of an alarm bell or a light, when there is a risk of any part of the body entering the danger zone. The guard should actually limit access to the danger zone, as shown in figure 12.

(c) *The guard should cause the operator no discomfort or inconvenience.* As already mentioned, guards which cause discomfort or inconvenience

Figure 12. Example of a guard effectively barring access to the danger zone. It is physically impossible for the operator's hands to be caught between the rollers

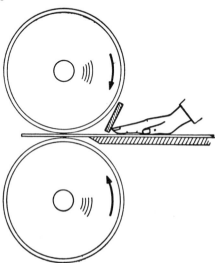

are put aside by the worker, and their usefulness is lost. An arrangement combining safety with ease of operation is shown in figure 13.

Figure 13. Reamer guarded by plexiglass screen, which does not obstruct the worker's view of the job

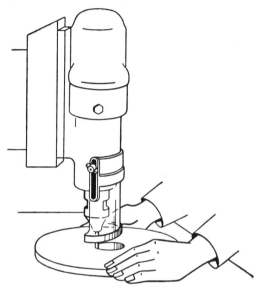

(d) *The guard should not interfere unnecessarily with production.* For this reason, guards such as two-hand systems on metal presses, or so-called "automatic" hoods for circular saws, should be avoided if other systems exist which give better protection and do not interfere with production. However, when it is impossible to find a guard that does not interfere with production, safety should take precedence over considerations of production, and imperfect protection should be preferred to an unprotected machine.

(e) *The guard should operate automatically or with minimum effort.* An example of an automatically operating guard is the hood for the cutter cylinder of a textile-shearing machine. This hood is connected with the starting mechanism in such a way that the hood cannot be opened when the machine is running, and when the hood is open the machine cannot start.

One type of guard for wood-planing machines, which has been in use for many years, consists of a screen placed above the machine shaft, pivoting on a vertical spindle placed at the side of the machine. The guard opens when the wood on the machine table touches it, and closes when the wood has passed the machine shaft. Such guards are often called "automatic guards", but they cannot be said to work automatically. They are completely unsatisfactory because they also open when a hand accidentally touches the screen and thus fail to give protection at the moment when it is most needed. Such a guard does not operate automatically at the critical moment.

One special type of automatic guard is the electronic guard which works with photo-electric cells. In this system, parallel light rays are projected in front of the danger zone of a machine. Interruption of these rays stops the machine or prevents it from starting. Such presence-sensing systems usually have a high sensitivity, and it is an advantage that there are no movable parts in front of the worker. However, special care has to be taken to ensure that the beam is wide enough, and that it is placed in such a way that no access to the danger zone is possible during operation.

(f) *The guard should be suitable for the job and the machine.* Too often, guards have been constructed which, while giving real protection, are not at all suitable for the job and are consequently not used.

Example. A sewing-machine factory designed a guard to prevent fingers from being pierced by the descending needle. The danger zone was perfectly protected, but the guard made it very difficult to thread the needle, and normal control of the work was impossible because the operator could not see what happened under the needle. This guard eventually had to be replaced by another which, besides giving adequate protection, allowed the needle to be threaded and the work to be controlled easily.

(g) *The guard should preferably constitute a built-in feature.* Much better results are usually obtained when a guard is part of the machine design rather than a later addition to the machine.

Example. The small hand- or electrically-operated meat mincer, used in factories and also in households, has a dangerous nipping point between the endless screw on the machine shaft and the casing near the feed opening. The various guards used to eliminate this hazard were often a nuisance to the worker, either during normal work or when cleaning the machine. A safe construction has now been designed. The feed opening is long enough but not too wide; hence, the fingers cannot reach the dangerous nipping point, and normal work and cleaning can be done without difficulty.

(h) *The guard should provide for machine oiling, inspection, adjustment and repair.* Where these requirements are not observed, it is necessary to remove the guard for each of these operations; in practice, it is often not replaced, with the result that the machine is unguarded when next used. Such difficulties have been encountered, in particular, with guards for power-transmission equipment.

Example. A machine pulley near the floor, with a belt running on an overhead pulley of the transmission shaft, should be protected by a steel angle frame, at least 1.9 m (6 ft) in height, with panels filled by perforated steel plates. For oiling and inspection, the guard should be provided with a small door with inclined hinges near the pulley. Such a door closes by gravity when left open. In this way, access to the pulley is easy and, at the same time, protection of the dangerous parts is guaranteed more or less automatically.

Figures 14 and 15 represent two types of non-removable guards designed to permit oiling and other maintenance operations to be performed without difficulty.

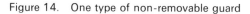

Figure 14. One type of non-removable guard

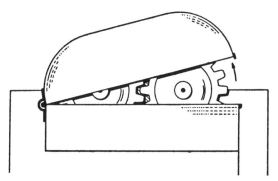

(i) and (j) *The guard should withstand long use with minimum maintenance and resist normal wear and shock.* It might seem unnecessary to mention this point particularly, since all guards should satisfy these requirements. However, many guards are constructed in a very flimsy way, either because some home-made device is considered adequate, or because insufficient allowance is made for ordinary wear and tear. Movable screens on metal presses have often proved defective because the fact that such screens may be opened and closed 800 times a day was overlooked in their design. The design of a guard calls for as much precision as the design of a machine, and satisfactory results cannot be expected without such precision.

Figure 15. Another type of non-removable guard

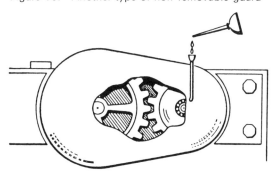

(k) *The guard should be durable, fire- and corrosion-resistant.* Here, special attention has to be paid to the choice of material used. If a guard is not durable, it will soon have to be replaced, and many examples show that, in such cases, replacement is not always made in time; the result is that the machine is used unguarded. Fire-resistant materials are always to be recommended, and corrosion-resistant materials are necessary if the guard comes into contact with chemicals or is used in very damp places.

(l) *Guards should not constitute a hazard by themselves (without splinters, sharp corners, rough edges or other sources of accidents).* There should not be any danger of the worker's fingers being nipped between moving guards on moving machine parts and a machine part.

Example. A machine for shearing metal was provided with a screen which was lowered automatically on to the table in front of the knives when the machine was started. Under normal conditions, the screen prevented a hand from entering the danger zone under the knives. However, if a hand was in the danger zone when the machine was started, the descending screen might press it on to the table and trap it in that position; the knives would then descend and cut off the fingers. A better arrangement would be for the knives to be blocked whenever the descending screen met some object.

(m) *The guard should protect against unforeseen operational contingencies, not merely against normally expected hazards.* Often, a machine is considered to be adequately guarded when no special risk exists under normal working conditions. Experience has shown that this assumption does not prevent accidents.

The example of the two-hand system for metal presses, previously mentioned in *(d)*, should also be considered here. The system consists of two handles, or two push-buttons, arranged in such a way that, in order to start the press, the worker has to push both down, and thus has to remove both hands from the danger zone of the press before the ram descends. However, apart from interfering with productive efficiency, this system gives no protection to any other person besides the operator (for example, to a supervisor who has been called to verify the working of the press).

Not all these requirements for guards can always be satisfied, but, as far as circumstances permit, they should be. It will be clear that designing a guard calls for more research and experience than the average employer can be expected to provide. It is not surprising, therefore, that, in some countries, the State has taken a hand in the designing of guards and, by drawing on all the inventive resources of the country, has provided industry with guards which are usually far superior to any home-made device.

GUARDS AND PRODUCTION COSTS

A suitable guard may not only give adequate protection, but may also improve the quality and quantity of the work done on the machine. Thus, a knowledge of the technical factors embodied in a particular machine is essential for any attempt to make it safe, and sometimes an efficient working method must be found before the designing of a guard can begin. This approach has proved to be particularly valuable for the protection of woodworking machinery, and it has also resulted in the designing of completely new systems for guarding machines such as metal presses and grinding wheels. The Swiss National Accident Insurance Institute was among the first to apply the principle that, firstly, an efficient way of operating the machine must be found and, secondly, the guard must facilitate operation as well as protect the operator.

Many employers (and, in particular, small employers) want to keep the cost of guards as low as possible. They often forget that guards which give incomplete protection may fail to prevent an accident, and thus may prove, in the end, much more expensive than an effective guard.

BUILT-IN GUARDS

There is now widespread support for the principle of "built-in safety"—that is to say, the principle that the most effective way of protecting a machine is to make the guard an integral part of the machine. Generally speaking, built-in guards are usually cheaper and more effective than guards added to the machine after it has been delivered to the user. Moreover, it is often much simpler for manufacturers to equip their machines with guards than for the individual user to do so. Many attempts have been made to make built-in safety for machines and other equipment a statutory obligation on manufacturers and distributors.[1] This obligation exists in varying forms in some countries, e.g. Austria, Denmark, France, the Netherlands, Sweden and the United Kingdom.

Though the requirement that machines should be guarded before they are delivered has its advantages, it also has its drawbacks. For instance, the relevant statutory provisions may be so strict as to make the construction of new safety appliances impossible.

Example. Some safety legislation prescribes that wood-planing machines must be equipped with round cutter blocks. This limits the freedom of construction. However, as the necessity for cutter blocks of other shapes has never been felt, the standardised round block has not given rise to difficulties.

Another problem is that the use that will be made of a machine is not always known to the manufacturer, while the type of guard required depends on this use.

Example. A metal press used exclusively for making objects from metal strips, as in the canning industry, should be guarded with a fixed enclosure of the ram to make the machine safe and ensure maximum production; when the machine is used for different types of work, an interlocked movable screen is necessary.

Finally, some guards are not fixed to the machine but to the floor, the wall or the ceiling near the machine.

Example. If, in a woodworking factory, very large workpieces have to be handled, it is necessary to keep the machine tables of circular saws and moulding machines free from any obstruction, and to hang the hood or other guard from the ceiling, to fix it on the wall or to place it on the floor near the machine; delivery of the machine with a guard attached to it would be impracticable.

Guards for transmission equipment, or guards required at the point of operation, can generally be produced by the manufacturer because, in these cases, developments have been consolidated and standardised. Calenders in laundries are one example. When no standard guard exists, it is impossible to do more than indicate that a certain hazard has to be eliminated.

In some cases, it is sufficient for safety regulations to prescribe that the machine should be so designed that it is possible to fit a suitable guard, without giving specifications for the guard itself. This is the case with the riving knife of a circular saw, for the specifications of the knife required depend on those of the saw blade to be used; it will not be difficult to guard the machine later if an adequate support for riving knives has been placed on the machine.

QUESTIONS

1. Give some reasons why workers may fail to use machine guards provided for them.

2. Mention some of the requirements that a machine guard should satisfy.

3. Describe the potential advantages and drawbacks of machine guarding.

4. What are the advantages of built-in guards?

Note

[1] See ILO: *Model code of safety regulations for industrial establishments*, Regulation 6, para. 2 (Geneva, 1949): "The builders, manufacturers or vendors shall comply with the provisions of this Code concerning the protection of machines, apparatus and vessels, and the packing and marking of flammable, explosive or toxic substances."

SOME PRINCIPLES OF
ACCIDENT PREVENTION
III. OTHER PROTECTIVE MEASURES

6

PLANNING

Good planning is as essential in safety as it is in production. If a new factory is to be built, or an existing factory reconstructed, there are many things affecting both safety and production that should be taken into account in the planning stage, such as the site, facilities for handling and storing materials and equipment, floors, lighting, heating, ventilation, lifts, boilers, pressure vessels, electrical installations, facilities for machinery maintenance and repair, and fire precautions.

It is essential that safety considerations be borne in mind at the time of the actual planning, and not as an afterthought when the factory has been built. Consequently, there should be a safety engineer in the planning team from start to finish.

The submission of plans for new and reconstructed factories to the labour inspectorate (or other competent authority) for comments or approval is also a useful precaution; indeed, in some countries it is compulsory. Good plans make for economy as well as safety. It is much cheaper to modify a plan than to alter a building.

Once a factory is in operation, planning is still essential in a number of fields to ensure the highest possible standards of safety as well as efficiency. Processes have to be organised, plant has to be arranged for them, methods of work have to be decided upon; changes will occur, from time to time, in the kinds of processes carried on; there may be minor structural alterations; new equipment will have to be acquired. All these things need planning. Here, again, better results are obtained more cheaply by planning safety measures beforehand than by improvising them afterwards.

The manager of a factory can generally follow a number of principles in planning for safe and efficient production. Here are some examples:

1. Keep the handling of materials and articles to the minimum.

2. Provide safe walking surfaces on floors, stairs, platforms, gangways, etc.

3. Provide adequate space for machinery and equipment.

4. Provide safe access to every place to which workers have to go.

5. Provide for the safety of maintenance and repair personnel, such as window-cleaners, and of men working on overhead equipment.

6. Provide safe transport facilities.

7. Provide adequate means of escape in case of fire.

8. Allow for expansion.

9. Isolate dangerous processes, such as spray-painting and processes with high fire or explosion risks.

10. If possible, only buy machines with built-in safety devices.

The following are some examples of measures that can be taken to reduce accident risks in the production process:

1. In the woodworking industry, tenoning can be done more safely on a moulding machine than on a circular saw; consequently, a moulding machine should be made available for such work.

2. In rubber-shoe factories, benzene-rubber solutions are used to glue different parts together. As benzene vapours, which can cause serious blood diseases, may escape into the factory premises, benzene should be replaced by a harmless, or less harmful, substitute product such as petrol, which, although about as flammable as benzene, is much less toxic.

3. In garages and car repair workshops, machine parts are often cleaned with petrol. The substitution of kerosene (paraffin) reduces fire risks considerably.

As already stated, the planning of repair and maintenance work is just as important, from the safety viewpoint, as the planning of layout and processes. The breakage of some machine part is often the cause, not only of an accident, but also of an interruption of work. Regular inspection, good maintenance and prompt repair will help considerably to improve the efficiency of work and reduce the number of accidents. The following are some examples:

1. In the chemical industry, inner baskets of hydro-extractors have sometimes burst because of corrosion caused by substances which have dried in them. Regular inspection of the extractors, and replacement of the corroded baskets in good time, could have prevented such accidents.

2. Accidents caused by broken chains or wires are often due to lack of regular inspection.

3. Many fatal accidents, caused by electricity, happen with electric hand tools. If they have been damaged in normal use, and have not been regularly examined, their casings can carry mains voltage. Thus, a worker who touches them in unfavourable conditions can be killed.

Particular care should be taken with the organisation of tool rooms. These rooms should have inspection and repair facilities to ensure that no dangerous tools (such as mushroomed chisels and defective electrical tools) are given out to workers. This also applies to other items of equipment such as chains, wire ropes and ladders.

Another important practical aspect of accident prevention is the worker's use of safe working methods. It is extremely difficult to change a person's way of thinking. In addition, as already indicated, it is very difficult, if not impossible, to make sure that a person who has corrected a wrong attitude does not become distracted by sorrow, illness or interests outside the factory and fall back, permanently or temporarily, into former habits.

The formation of safe working habits is a different matter; it involves the absorption of a safe working method until it has become second nature and is followed automatically. When this is so, one may expect that the same work method will be followed, whether the worker is thinking about the work or not; this should guarantee safe working in all circumstances. In addition, some accidents, due to unsafe acts, improper attitudes and the like, will be prevented.

The importance of teaching safe working habits to young persons is self-evident. Adult workers often have difficulty in changing working habits, and so once incorrect habits have been formed, it is difficult, but not impossible, to correct them.

Good working habits also include taking proper care of machines and tools, keeping them in good condition and, in particular, keeping cutting tools well shaped and sharp.

GOOD ORDER AND GOOD HOUSEKEEPING

A third group of safety measures consists of those related to good order and good housekeeping all over the plant. If there is a place for everything, and if everything is in its place, a considerable number of accidents are likely to be avoided.

Good order means, in the first place, the removal of objects which obstruct passageways; collisions and stumbling are prevented and it is easier to escape in emergencies. Passageways should be clearly marked off (e.g. with white lines) and should not be used for storage of materials. Good order also means that materials must be properly stored and waste materials promptly removed. Cotton waste, for instance, should be placed in closed metal containers — a measure that promotes good order and limits fire hazards at the same time.

Dry, clean, ventilated store-rooms with suitable racks, hooks, etc., should be provided for electric hand tools, other tools, chains, wire ropes, ladders, and so on; and, when not in use, these items should be kept in their proper places in the store-room. They should be inspected regularly and defective equipment should be discarded. Containers for flammable liquids should be completely airtight to prevent any leakage.

The following are examples of improvements that good housekeeping can bring about:

Accident prevention

1. If suitable containers for spilled and leaking oil are placed under the barrels containing lubricating oil in engine rooms, the floor will not become oily, and therefore slippery.

Figure 16.   The kind of accident that occurs when floors are allowed to become oily

2. The removal of vapours from textile-dyeing rooms will contribute not only to a healthier environment, but also to better visibility, and, in this way, to safety. It will also help to reduce the cost of repair and maintenance of the building.
3. In car repair workshops, the risks of tripping over tools and engine parts would be lessened if workers were provided with small trolleys, containing compartments and drawers for a complete set of tools, and spaces for small machine parts removed during repair work.

Good order and good housekeeping not only reduce accident risks by eliminating physical risks but also contribute to safety by their psychological effect. When much thought is given to good order, and when good housekeeping is the universal practice, a worker will most probably behave more carefully than when disorder prevails and housekeeping is neglected.

It will be clear that good order and good housekeeping can be achieved more easily if workers support the idea and obey all instructions designed to promote it—for instance, by keeping gangways free from obstructions, using receptacles for waste materials and storing tools in their proper places. Once good habits are acquired, it is not too difficult to preserve them, for good housekeeping not only helps to prevent accidents but also makes work easier. It would be interesting to know how much time is lost by looking for misplaced tools, or for seeking the right bolt, nut or washer, when such items are not kept in any particular place.

The saying, "safety pays", is true in every sense when applied to maintaining good order in workplaces.

WORKING CLOTHES

The *Model code of safety regulations for industrial establishments* contains an extremely comprehensive summary of safety requirements regarding working clothes, which reads as follows:

Regulation 226. Working clothes

1. When selecting working clothes, consideration should be given to the hazards to which the wearer may be exposed, and those types should be selected which will reduce the hazards to the minimum attainable in each case.
2. Working clothes should fit well; there should be no loose flaps or strings, and pockets, if any, should be few and as small as practicable.
3. Loose, torn or ragged garments, neckties and key chains or watch chains shall not be worn near moving parts of machines.
4. When the operations involve a danger of explosion or fire, it shall be prohibited, during working hours, to wear articles such as collars, eyeshades, cap visors and spectacle frames made of celluloid or other flammable materials.
5. Shirts with short sleeves should be worn in preference to shirts with rolled-up sleeves.
6. Sharp or pointed objects, explosive substances or flammable liquids shall not be carried in pockets.
7. Persons exposed to flammable, explosive or toxic dusts shall not wear clothing having pockets, cuffs or turn-ups that might collect such dusts.

Workers' clothes and footwear are often completely unsuitable for work in factories. Women, in particular, but sometimes also men, work in old clothes and shoes which they consider to be no longer good enough for street wear. In some countries, local custom obliges women to wear long veils, even when working near machines.

The wearing of clothes with loose attachments has caused many accidents.

Example. An experienced worker started to clean a pit under a machine. To protect himself from liquid dripping down from the machine, he put a cloth on his head, fixed it under the chin and left the loose end hanging over his shoulder. Some minutes later, he was found in the pit decapitated. The cloth had been caught by a revolving shaft, which was about 1.20 m (4 ft) above the bottom of the pit.

Example. In a dairy-products plant, a painter was working on a ladder near a revolving transmission shaft. Suddenly, his sleeve was caught by the shaft, with the result that his clothes were torn to pieces and he suffered several injuries.

If working clothes are exposed to heavy wear and tear, moisture or dirt, suitable types of clothing should be provided; in most countries, the workers have to buy their own clothing if it is not provided for them.

Veils, and other clothes with loose parts, which are worn because of national custom or religious opinion, may present a problem. The only way to eliminate undesirable risks in such cases may be to prohibit the employment of such persons in types of work in which clothes are liable to be caught in moving machine parts.

Ordinary working clothes cannot protect the wearer against hot metal, acids, flying fragments and various other risks; here, personal protective equipment is required.

The wearing of finger rings during working hours has often resulted in the loss of a finger when a ring has been caught in moving machinery.

Falls on the same level (i.e. tripping, slipping, etc.) are among the commonest of all accidents; many of them are doubtless due to unsuitable footwear. The high heels of women's shoes are particularly dangerous on factory floors; they are liable to cause falls as a result of poor balance, slipping, or becoming caught in gratings or small irregularities. Falls may also be caused by loose laces and ragged soles or heels.

PERSONAL PROTECTIVE EQUIPMENT

The best way of preventing accidents is undoubtedly by eliminating the hazard, or controlling it as close to the source as possible; but if this is impossible it may become necessary to provide the worker with some sort of protective clothing.

Figure 17. Typical items of personal protective equipment

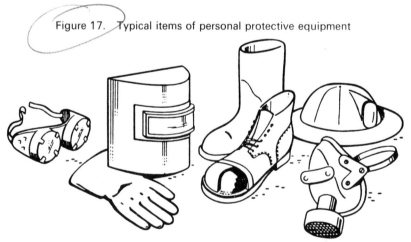

Personal protection should be considered as the last line of defence, because often the equipment is cumbersome to wear and restricts movement. Therefore, not surprisingly, it is sometimes discarded by workers. Just as the equipment has been designed to prevent outside hazards from affecting the body of the worker, so it traps body temperature and water vapour inside, with the result that the worker becomes hot and sweaty and tires more easily.

In many countries, standards for personal protective equipment have been drawn up by national standard institutes. However, this in itself can lead to problems. For example, a respirator that fits a large, clean-shaven worker is unlikely to fit a small bearded colleague; and yet, so often, the stores contain only one type of standard mask. Although this may be an extreme example, it illustrates the point. All personal protective equipment must, as the name implies, fit the worker in question, otherwise its value is purely cosmetic and the worker is given a false sense of security. In the case of the respirator mentioned above, the term "fit" means that there is an effective seal between the face and the mask and that there are no leaks.

A number of basic criteria must be met by all types of protective equipment. Probably the two most important are:

(a) whatever the nature of the hazard, the equipment or clothing must provide adequate protection against the specific hazard; and

(b) such equipment or clothing must be light to wear and durable, and should cause the minimum of discomfort but allow maximum mobility, visibility, etc.

Eye protection

It is important to ensure not only that protective equipment is worn, but that it is worn correctly. For example, workers not accustomed to wearing glasses sometimes reject the various types of eye protective equipment because it is a nuisance and causes discomfort. Workers must be told the reasons for wearing the device, and it should be explained that no safer alternative exists. Often, workers who think that the risk of an eye accident is a very real one will wear safety glasses willingly, but those who think the risk very small will object.

This difficulty can be overcome in different ways. In some undertakings, areas where there is thought to be a real risk of eye accidents may be entered only if glasses are worn. Consequently, in these places, all workers wear glasses during the whole of working time, and it is accepted as normal practice.

Other factories make a large number of different types of glasses available, and workers can make their own choice; correct use of the glasses is ensured by regular inspection. In such cases, workers are not obliged to wear glasses of types which the management considers suitable but they do not.

Some workers may not find suitable glasses because they have eye defects. It is therefore desirable that the management should arrange for the workers' eyes to be tested and for advice to be given on the most suitable type of safety glasses for the job, the eye distance, etc. If necessary, advice can be given on the fitting of prescription glasses at the same time.

Safety shoes

Safety shoes should protect workers against accidents caused by heavy objects dropping on the feet, protruding nails, molten metal, acids, etc. Ordinary leather shoes in good condition give some protection against crushing and piercing, but, to be really safe, shoes with built-in steel toe-caps, and with steel soles inside the leather ones, should be worn. The latter precaution is particularly important for workers on building sites, where protruding nails may be common hazards.

Sometimes, special kinds of footwear are necessary. For instance, electricians should wear non-conducting shoes (i.e. shoes with no metal nails), and workers in explosives factories should wear non-sparking shoes (also without metal nails).

Gloves

Gloves should not only protect workers from hazards but should also allow the fingers and hands to move freely. The kind of glove required will vary according to the injury to be prevented (puncture, cut, heat burn, chemical burn, electric shock, radiation burn, etc.). It should be remembered that it is dangerous to wear gloves when working at drilling machines, power presses and other machines in which a glove might be caught.

Hard hats

Workers liable to be struck by falling or flying objects, or otherwise exposed to head injuries, should wear hard hats or helmets, which are strong enough to protect them, but not too heavy. Plastic hard hats have proved to be very suitable. Common practice now is to designate any area where there is the possibility of such accidents as a "hard hat area". This simply means that anyone entering the area must put on a hard hat.

Other protective equipment

Protective equipment exists for almost every type of hazard and for all situations. Such equipment ranges in complexity from a simple pair of gloves to pressurised whole-body suits. During the past few years, much progress has been made in many countries in the design of suitable personal protective equipment, and the number of glasses, shoes, helmets, and other items used in different branches of industry has increased considerably. However, there are still countries where personal protective equipment is hardly known and where, consequently, workers are unnecessarily exposed to risks.

However, it must be remembered that asking a worker to wear protective clothing is an admission that a hazard exists and that it cannot be controlled by better methods. As was stated earlier, the use of personal protective equipment must be considered as the last line of defence.

COLOURS, NOTICES, SIGNS, LABELS

Colours

Colours may be used for a variety of purposes in the interests of safety, as the following examples will show:

(a) general safety colour codes are used to identify danger spots, fire-protection equipment, first-aid equipment, exits, traffic lanes, and so on;

(b) special colour codes are used to identify the contents of gas cylinders and piping;

(c) suitable colour schemes can improve perception and visibility in workrooms, passageways, etc.; and

(d) attractive colour schemes for walls, ceilings, equipment, etc., can have a good psychological effect.

Various colour codes have been used over the years. A recommendation made by the International Organisation for Standardisation (ISO) defines the meaning and use of certain safety colours: amber (orange-yellow) is used to indicate danger (for example, to identify places normally covered by guards so that it is easy to see when the guard is missing); red is used for stop signals, emergency stop devices and fire-fighting equipment; and green is used for escape routes, first-aid stations, "go" traffic signals and safety installations generally.

Columns or other fixed objects near passageways, and obstructions of all kinds, are often painted with inclined yellow and black lines, which contrast well with their surroundings; traffic lanes may be marked with white lines.

Accident prevention

Notices and signs

Notices and signs can also serve a variety of purposes. They can convey instructions, warnings or general information. They are not a general substitute for protective measures and safety instructions, but can be a very useful supplement to them.

"No Smoking" is one of the commonest examples of an instructional notice; it is a reminder of the danger of smoking in places with a fire risk. Another common use of notices is to prohibit the opening of locked valves and switches controlling equipment on which maintenance work is being done. Quite a number of notices and signs can be used to regulate transport in the factory.

"High Voltage", "Level Crossing" and "Caution. Men at Work" are other examples of warning notices.

Informational notices serve to indicate the whereabouts of exits, first-aid posts, etc.

If notices and signs are used too much, they may lead to confusion and will probably be ignored.

Labels

Dangerous substances, and the containers for such substances, should be properly labelled. Many accidents occur because toxic, corrosive, flammable or other dangerous substances are kept in containers that do not show that the contents are dangerous, or — even worse — in ordinary drink containers. Very serious accidents have occurred when workers have drunk poisons kept in milk or beer bottles.

Example. A worker, seeing a beer bottle near the place of a fellow worker who had gone away for some minutes, took the bottle and, to play a joke on his mate, drank the contents. The bottle was in fact filled with mordant, and the worker had to be taken to hospital immediately; he was treated there for two weeks.

As an aid to the prevention of such accidents, suitable labels, such as those given in figures 18 to 23, should be used. The symbols were originally designed by a meeting of experts on dangerous substances, convened by the ILO in 1956 to draw attention to the risks associated with dangerous substances.

The use of symbols for this purpose has the advantage that the labels can be understood by illiterate persons. However, it is desirable to add to such a symbol a text indicating the following:

(a) the name of the substances;

(b) a description of the main risk, or risks;

(c) a statement of the main precautions to be taken; and,

(d) if necessary, an indication of first-aid or other simple measures to be taken in the case of injury or emergency.

Figures 18-23. Symbols for labelling of dangerous substances

Figure 18. Toxic substances

Figure 19. Explosive substances

Figure 20. Flammable substances

Figure 21. Oxidising substances

Figure 22. Corrosive substances

Figure 23. Radioactive substances[1]

Figures 24 and 25.   Examples of labels used for dangerous substances

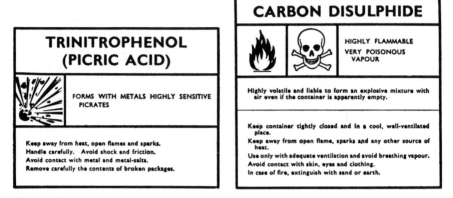

Figure 24.   Picric acid   Figure 25.   Carbon disulphide

LIGHTING

Lighting is important as a safety factor in the physical environment of the worker. Several investigations into the relation between production and lighting have shown that adequate lighting, arranged to suit the type of work to be done, may result in a maximum of production and a minimum of inefficiency, and thereby may help indirectly to reduce the number of accidents. To the extent that accidents result from visual fatigue, adequate lighting is a preventive measure; and the relation between poor lighting and high accident rates has been demonstrated in a number of publications for many years.

Lighting factors that contribute to accidents include direct glare, reflected glare from the work, and dark shadows; suddenly passing from bright surroundings into darkness, or vice versa, may also be dangerous. Sometimes, also, what appears to be carelessness may be the result of difficulty in seeing.

The following are two examples of accidents caused by abrupt transition from bright light to darkness:

Example. During the night, materials were being stored with a crane. The place where they were stored was lighted with a shaded acetylene lamp of high intensity. The lampshade caused a strong contrast between the lighted area and the adjacent areas, which were in complete darkness. Suddenly, the crane slewed and a man walking in the dark sector was hit by the load and seriously injured. It was impossible for the crane driver, whose eyes were accustomed to the light of the acetylene lamp, to adapt his vision fast enough to see what was going on outside the lighted area, and avoid the accident.

Example. Poor lighting threw sharp shadows on the treads of some stairs. A man descending the staircase missed his footing and fell.

Adequate lighting is particularly important for accident prevention in places where there is a risk of stumbling or falling (e.g. near quaysides, railway lines, etc., and in gangways, on staircases and near exits which have to be used in emergencies).

When there are large numbers of persons in workrooms, it is essential for gangways, staircases and exits (and, if necessary, places near dangerous machines) to be kept lighted under all circumstances, even if normal lighting fails. In practice, this is a difficult problem. Small electric generators, which can feed a group of lamps in an emergency independently of the normal supply of electricity, are only available in a limited number of countries, and even then they are not widely used. Another solution is to use candles or kerosene lamps at strategic points in gangways and on staircases, but they must be lit every evening, and this is seldom done regularly. In addition, such lights may give rise to fire hazards in certain types of factories. In some cases, lines and arrows have been painted on floors and walls with luminous paints so that the exits can be seen if normal light fails. Light signs indicating emergency exits and placed near the ceiling are not always suitable, as, in the case of fire, they may be hidden by smoke.

VENTILATION AND TEMPERATURE CONTROL

Ventilation, whether general ventilation or local exhaust ventilation, falls mainly within the province of industrial hygiene, but is of some importance from the safety standpoint. The same is true of air conditioning. Exhaust ventilation, for example, is one means of removing explosive dusts, such as those of aluminium, magnesium, cork, starch and flour, from the working atmosphere. Flammable vapours in the atmosphere can be diluted to their so-called safe limits by general ventilation, or can be removed altogether by exhaust ventilation. Air conditioning can prevent excessive cold and excessive heat, both of which have been found to be conducive to accidents.

Ventilation systems, however, require careful designing – this applies particularly to exhaust ventilation systems, which, if badly designed, can be worse than no ventilation at all. Exhaust hoods, or slots, should be so located that no part of the fumes or dust being removed can enter the worker's breathing zone (see figure 26).

NOISE

There is no hard-and-fast definition of excessive noise, but there is a wide measure of agreement that any sound intensity above 90 dB(A) is objectionable to workers, and that high-pitched sounds may be

Figure 26. Well designed (left) and badly designed (right) exhaust ventilation systems

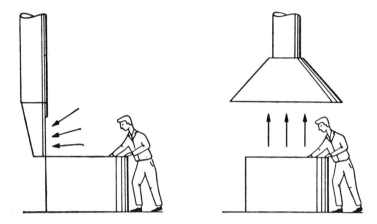

objectionable at lower intensities. Intensities above 90 dB(A) are commonly encountered in riveting operations, circular sawing, weaving sheds, boiler shops and aircraft-engine testing shops.

Excessive noise makes communication between workers very difficult, makes hearing of warning signals impossible, causes misunderstandings and leads to permanent loss of hearing. For these reasons, noise must not be overlooked in safety-planning. In addition, noise can be extremely tiring, and, in this respect, it has the same ill-effects as other types of fatigue.

Safety precautions against noise take the form of specially designed machines to prevent noise, measures to absorb noise and vibrations, and, where absorption is impracticable, limiting the number of persons exposed to excessive noise as far as possible. Only as a last resort should ear plugs, ear muffs, helmets or other specialised ear protectors be used.

QUESTIONS

   1. Mention some ways in which planning can improve safety standards.

   2. What is the value of good working habits?

   3. Give some examples of good housekeeping.

   4. What kinds of working clothes are dangerous —

(a) around machinery?

(b) near fires?

(c) in explosive atmospheres?

5. What sorts of injuries might be avoided by the wearing of—

(a) hard hats?

(b) safety boots?

(c) safety glasses?

(d) respirators?

6. Do you think all workers should wear safety glasses? Give reasons for your answer.

7. Mention some of the ways in which colours can be used for accident prevention.

8. How can labels help to prevent accidents?

9. Mention some of the requirements that good lighting should satisfy.

10. What bearing has ventilation on safety?

11. How can noise be dangerous?

Note

[1] The skull and crossbones do not appear on labels intended for the transport of parcels containing radioactive sources other than large sources as defined in the International Atomic Energy Agency's *Regulations for the safe transport of radioactive materials.*

SOME PRACTICAL APPLICATIONS OF ACCIDENT PREVENTION PRINCIPLES

7

It has become apparent from the preceding lessons that many volumes could be written about the safety measures desirable in the planning of various parts of a factory building, the designing and installing of equipment that may be housed there, and the handling and storage of the dangerous substances that may be used or manufactured in it. It is not possible in this manual to explain in any detail the precautions that should be taken against all the risks that may arise in the factory building itself, in the equipment, or in the processes carried on in the factory. However, it is particularly important that every worker should know the main precautions that can be taken with certain items of equipment in general use (such as hand tools, portable electrical apparatus, gears, silos, acetylene cylinders and ladders). These precautions are explained in this lesson. Many of the general principles referred to in this section are as relevant today as they always have been, and should not be dismissed as trivial. For convenience, reference is made at times to the relevant provisions of the *Model code of safety regulations for industrial establishments*, which was mentioned earlier.

HAND TOOLS

A large number of injuries are caused by hand tools. Most of them are not very serious, but long periods of absence from work may be necessary if the injury becomes infected. The main causes of these accidents are the use of unsuitable tools, the misuse of tools, poor maintenance and improper storage. Regulations 119 and 211 (the latter dealing with the more specific subject of tools for maintenance and repair work) of the *Model code* contain general recommendations for the prevention of such accidents; these are quoted below with explanatory comments.

Accident prevention

Regulation 119

1. *Hand tools for factory use shall be of material of good quality and appropriate for the work for which they will be used.* Many accidents caused by breakages of tools, or parts of tools (such as handles), occur because the material is of poor quality. Tools in general should be made of best-quality steel, and handles for hammers, axes and similar tools should be made of good-quality hickory, ash or maple. In many countries, recommendations on the quality of materials for tools are laid down in national standards. As, in practice, it is often difficult to verify the quality of materials, tools should be purchased from specialised firms of high repute. Tools unsuitable for the work to be done, such as hammers of incorrect shape or weight, and wrenches which are too long or too short, should not be used. National standards in different countries deal with these points too. In the factory itself, it is the responsibility of the tool-buyer to see that the right tools are available, and it is the responsibility of the worker to use them in the right way.

2. *Hand tools shall be used only for the specific purposes for which they were designed.* The use of hand tools for purposes other than those for which they were designed (for instance, using a knife as a screwdriver or a spanner as a hammer) is dangerous because the tool may break, splinter or slip, and so cause an accident.

Figure 27.   Never do this. Use a hammer instead

Figure 28.   This is *dangerous.* Find a wrench of the proper size

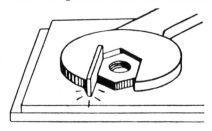

Tools are often misused because the right one is not available when it is wanted. If a worker wants a wrench that is missing, he or she will not want to stop work to look for it, but will use one with an opening too large for the nut and put in filler plates. It is easy to forget not only that a wrench so used may slip off the nut, but also that it is too long for the nut and that, if too much pressure is applied, the bolt may snap — an accident which in certain circumstances might result in injury to the worker — or be otherwise damaged, with a resultant loss of time.

The remedy is careful planning of the composition of sets of tools for different workers — possibly in consultation with the workers themselves — and regular inspection to check that no tools are missing. All workers who know their jobs will certainly use the right tools if they are provided.

3. *Wooden handles of hand tools shall be —*

(a) *best-quality, straight-grained material;*

(b) *of suitable shape and size; and*

(c) *smooth, without splinters or sharp edges.*

A good-quality, springy wood, such as hickory, ash or maple, should be used for handles. The length of the handle will obviously depend on the type of tool (hammer, axe) for which it is used. The handle should be shaped to fit the opening of the head of the hammer or axe. The joint between the head and the handle should be secured by a wedge, preferably of hard wood. Special care should be taken to fix the head of a hammer (see figures 29 and 31) at right angles to the handle, otherwise the joint will be dangerously weakened, and, when the hammer is used, the head may fly up and injure somebody.

Figure 29.   What is liable to happen if the head of a hammer is not fixed to the shaft at right angles

4. *Where there is any risk of an explosive atmosphere being ignited by sparks, any hand tools used therein shall be of a non-sparking type.* Explosive atmospheres can be expected in places where flammable liquids

are manufactured, handled or stored, e.g. in places where petrol is stored or calcium carbide containers have to be opened. In such cases, the tools used should be made of wood, hard rubber, copper, beryllium alloy or some other alloy which does not produce sparks when hit.

Figure 30. If there are flammable vapours about, do not use chisels (like this one) which produce sparks when hit

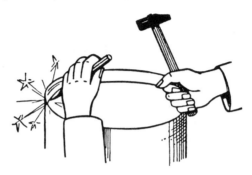

5. *Hammers and sledges, cold chisels, cutters, punches and other similar shock tools should be made of carefully selected steel, hard enough to withstand blows without mushrooming extensively but not so hard as to chip or break.* The hardness of steel shock tools has a great influence on safety. Excessively soft steel will soon mushroom at the point of impact. If the tool is not ground in time, small particles can be loosened by a blow and fly off, endangering persons (their eyes in particular) in the neighbourhood. Excessively hard steel, too, may splinter on impact; such splinters sometimes penetrate very deeply into the eye, and sight may be entirely lost. To eliminate these risks, steel used for shock tools must be neither too soft nor too hard; the limits are often laid down.

6. *Heads of shock tools should be dressed or ground to a suitable radius on the edge as soon as they begin to mushroom or crack.* The risk of flying splinters can be reduced by rounding the edges of hammers, anvils, etc. (see figure 31).

Figure 31.

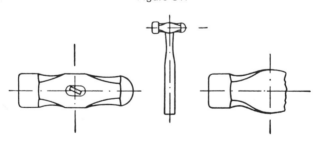

7. *Hand tools should be tempered, dressed and repaired only by properly qualified persons.* As the properties of steel depend on the way in which it has been hardened and tempered, this work should be done only by competent persons who are fully aware of the consequences of changes in temperature or treatment. Repair work should also be done only by qualified persons, in order to prevent damage to the tools or to the materials from which they are made.

8. *When not in use, sharp-edged or sharp-pointed hand tools shall be provided with protection for the edges or points.* Protection of the sharp edges and points of tools prevents injury by accidental contact, and at the same time prevents tools from being damaged if they are dropped on stone floors or on other solid objects. Leather sheaths or hoods may be used for such protection.

9. *Hand tools shall not be allowed to lie on floors, passageways, stairways, or in other places where persons have to work or pass, or on elevations from which they may fall on persons below.* This provision is necessary to prevent accidents caused by workers' tripping over objects or by falling objects. However, full compliance can be expected only if suitable containers are available for those tools not in use during work. Workers should be provided with tool boxes, small trolleys or other convenient means of storing tools. A simple precaution to prevent tools from falling off step-ladders is to make a slot, in which tools can be placed temporarily, in the highest tread of the ladder.

10. *Suitable and conveniently located cabinets, holders or shelves shall be provided at benches or machines for hand tools.* Persons working at benches or machines should also be provided with proper facilities for storing tools. It is important that there should be sufficient space for every tool, and that the tools can be stored in such a way that every one can be found, and checked, easily. The general rule for good housekeeping — "for everything a place and everything in its place" — should be strictly applied.

11. *Hand tools should be —*

(a) *issued through a tool room, in which they are stored safely on racks or shelves in cabinets or tool boxes;*

(b) *inspected periodically by competent persons; and*

(c) *replaced or repaired when found defective.*

It is easier to inspect hand tools if they are kept in a tool room and distributed by storekeepers. However, it is sometimes more convenient to leave workers permanently in charge of some tools, and in other cases it is customary for workers to have their own. In such cases, it would be desirable to make a rule that all tools should be handed in to the tool room periodically for inspection and examination by competent persons, to ensure that they are properly looked after and in good condition. Tool-room attendants should have strict instructions not to issue damaged or otherwise unsuitable tools.

Figure 32.   These are *dangerous* and should be *scrapped*

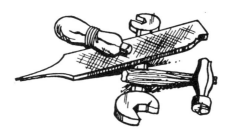

12. *Workers should be properly instructed and trained in the safe use of their hand tools.* Training has an indirect influence on safety standards, as insufficient skill often leads to improper use. Regular inspection of work habits offers an opportunity to discover where additional training and instruction is necessary.

Figure 33.   Incorrect and correct way of using a screwdriver

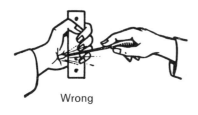

Wrong

Right

Regulation 211

(1) . . . (8)

9. *A sufficient supply of tools of the various types required shall always be kept available for maintenance and repair men.* A sufficient supply of tools is necessary not only for the reasons already mentioned under Regulation 119, paragraph 2, but also as a guarantee that maintenance and repair work will be done correctly, thus eliminating the risks due to defective repair work.

10. *Such tools shall be kept in safe condition and should be inspected at regular intervals by a competent person appointed by the management.* The observations on Regulation 199, paragraph 11, apply here also.

11. (1) *Maintenance and repair personnel shall be provided with special tool bags or portable tool boxes of a size sufficient to hold all the hand tools needed for their work and so constructed that they can easily and safely be hoisted on to platforms and other elevated workplaces.*

(2) *Where necessary, special hand trucks shall be provided for the transport of heavy tools needed in repair and maintenance work.* As maintenance and repair people may have to go anywhere in the factory, they should be provided with tool boxes, or tool trolleys, in which the tools can be fixed in such a way that they will not be disturbed even if the box or trolley is tilted. One way of keeping the tools secure is to enclose boards or rubber sheets, with openings into which the different tools fit, beneath a cover. Boxes and trolleys should be so constructed that they can be moved about without difficulty.

12. *In large establishments special fixed tool cabinets or tool boxes for maintenance and repair men should be provided in each department, particularly where special tools or tools too heavy to carry over considerable distances may be frequently needed for their work.* Unnecessary movement of tools should be avoided. As maintenance and repair work may be needed urgently to prevent serious damage to factory equipment or disruption of production, or to avert accidents, the tools necessary for this work should be immediately available at all times.

13. *All repair men should be provided with strong electric flashlights, which shall preferably be of the flame-proof type.* Because of fire risks, repair men should not use matches or candles to see by in dark places. Flame-proof flashlights are essential if highly flammable substances are present (e.g. in textile factories and flour mills).

PORTABLE ELECTRICAL APPARATUS

Many accidents occur because the casings of portable electrical apparatus (such as power drills) are carrying mains voltage as a result of some defect inside the apparatus. Accidents have also been caused by defects in the socket, the plug or the flexible cable. When a person touches an electrically charged object, there may be a serious, or even fatal, accident if the voltage is above a certain threshold.[1] As the mains voltage of electrical installations is higher than this threshold, workers must always take special precautions when using portable electrical apparatus.

One essential measure to prevent accidents of this kind is the earthing of casings. This can be done by having an extra wire, inside the cable, attached to a third pin in the plug which connects with a good earth through the socket. This method is not, however, completely reliable because the electric cable of portable apparatus is often carried or dragged about, or laid on floors where it is liable to be trodden on, with the result that sometimes a wire breaks. If the earth wire is the one affected, the break is often not discovered immediately; and if, subsequently, an internal fault develops in the appliance, the casing may become live, and the next worker to use the appliance may receive a severe shock. The broken earth wire may even touch one of the live wires and cause the casing to carry mains voltage. It is safer, therefore, not only to attach the earth wire as described above, but also to attach one end of a loose wire to the casing of the apparatus and the other to an earthed point. This type of earthing is more reliable, but its efficiency depends on the worker, since he or she is responsible for attaching the second earth wire before plugging in the apparatus.

Figure 34 shows an electrical hand drill with a socket outlet well earthed by being joined to a fixed earth conductor or to an earth pipe, e.g. the main water supply pipe (gas piping should never be used as an earth). The earthing system is connected to a special socket contact. This socket contact receives a pin, which, in turn, is connected to a special earth wire in the cable. Consequently, the cable contains three wires; two are current-carrying, and the third earths the drill casing. The drill is provided with a switch, which is self-releasing. If, for greater safety, a second, visible earth conductor is used, this should be connected to the earth pipe before the plug of the drill is connected.

Figure 34.   The earthing system for an electrical hand drill

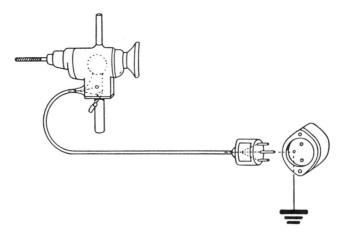

Another system that prevents serious accidents is the use of apparatus working on a current of low voltage, and of transformers that reduce the mains voltage to the requisite level.

Such precautions are not necessary when the casing of the tool is made of insulating material and when protruding metal parts (shaft, etc.) are insulated from the motor (e.g. by a coupling made of insulating material). Nor are special precautions necessary if the apparatus has double insulation — that is, if it is so constructed that not only are the different parts normally insulated, but also the casing is lined with insulating material and the protruding parts are constructed as described above.

When tools working at very low voltages are not available, safety can be increased by using an isolating transformer rated at 220/110 or 110/110 volts (see figure 35). The tool is operated, not by current direct from the mains, but by current from a completely separate secondary circuit set up by the transformer; in the event of current leaking into the casing, the worker will not receive the full shock of the mains current. Since the secondary circuit is completely isolated from the earth, there is less danger of touching a live casing.

Figure 35. An electric power drill with an isolating transformer

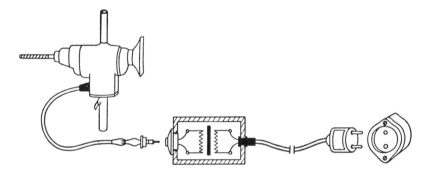

In this case, it is not necessary to earth the portable apparatus. However, as the safety of this system is based on the fact that the tool circuit is completely isolated from earth, care should be taken to ensure that the cable of the secondary circuit is in good condition. It is advisable to have this cable as short as possible. If necessary, a long primary cable should be used, and the transformer should be placed as near as possible to the point of operation. However, the transformer should never be put in a place where the use of voltages higher than the threshold voltages mentioned earlier is particularly dangerous (e.g. in boilers, double bottoms of ships, etc.). The midpoint of the secondary winding of the transformer should not be earthed.

To ensure that the apparatus is not used without the transformer, the plug at the end of the primary cable, and the one used for connecting the apparatus to the secondary cable, should be of completely different types.

Moreover, there should be a separate transformer for every appliance.

Every transformer used for safety purposes should have the primary and secondary windings completely separated, so that, even in the case of a defect, the two cannot come into contact.

Experience has shown that proper maintenance of portable tools, and daily inspection after working hours, can reduce the number of accidents considerably.

Lastly, attention should be drawn to a dangerous aspect of portable electric apparatus that can easily be overlooked. Some equipment has been designed and manufactured in such a way that a small metal portion has been left unearthed (e.g. a metal switch recessed into an insulated handle); it is therefore liable to become electrically charged, with consequent danger to the user. Workers using apparatus of this kind should be particularly careful.

GEARS

Regulation 76 of the *Model code* makes the following recommendations for the safeguarding of gears:

1. *Exposed power-driven gears shall be guarded in one of the following ways:*

(a) *with a complete enclosure;*

(b) *if the gear wheels are of the solid disc type, with a band guard covering the face of the gear and having flanges extending inward beyond the root of the teeth on the exposed side or sides.*

2. *Hand-operated gears shall be guarded in a manner similar to that prescribed for power-driven gears whenever they present a hazard.*

Generally speaking, the danger inherent in gears can only be eliminated by enclosing them completely. The guard should be constructed in such a way that it can be oiled (e.g. through small openings) without being removed. An example was given earlier (see figure 15).

If this is not practicable, or if the gear wheels have to be replaced from time to time, the guard should be designed so that it locks when the machine is running. This system is used for some textile machines, for instance. When such an arrangement is impracticable, the guard should be placed so that it inconveniences the worker when it is open, and has to be closed if the worker wants to work comfortably.

An example of a guard for gears was shown in figure 14. When the gears have to be changed in accordance with the work to be done, the guard can pivot on the bolt. If the gears were completely enclosed, it would be necessary to take the guard off each time the gears had to be changed, and it is doubtful whether, in practice, such a guard would always be replaced.

A guard that covers only the area where two gear wheels meet is inadequate, because it does not prevent a finger from penetrating between the gear wheels from a direction parallel to the shaft, because there is a risk of fingers being nipped between the end of the guard and the teeth of the wheel, and because the guard is so made that a small incidental deformation brings it close to the teeth and creates a similar risk.

The risk of fingers being nipped between two gear wheels is not their only danger. Gears present the same risks as other protruding parts of a revolving shaft — they can catch loose clothes, or materials such as fibres, yarns, cords, and so on. Spoked gears may cause accidents if a hand or some object penetrates among the spokes.

The hazards of gears on machines turned by hand are often underestimated, as may be seen from the following example.

Example. In a small bakery, a hand-driven machine was used for cutting dough in the shape of biscuits. As this machine was seldom used, was hand-driven, and was designed to be used by one person, and as only a father and his grown-up son worked in the bakery, guarding the gears of the machine was thought to be unnecessary.

One day, the machine was being used by the father and son together, the father turning the handwheel and the son feeding the machine with dough. Suddenly, the son's hand caught between the gears, and, before he could shout to his father to stop turning the wheel, he had lost the tips of two fingers.

Accidents like this show that it is essential to enclose gears completely, whether they are slow-running or fast-running, and whether they are hand-driven or power-driven.

LARGE CONTAINERS (SILOS, BINS) FOR DRY BULK MATERIALS[2]

Fatal accidents have occurred in hoppers and bins containing dry bulk materials when persons have entered them to try to start the material moving after it had stopped for one reason or another.

The following is typical of the kind of accident which may occur. Under certain circumstances, the material stored in a bin or silo is liable to stop flowing out, even though the outlet funnel at the bottom is open. Sometimes, this is due to the construction (the unsuitable shape of the bin and of the bottom in particular) or to the unsuitability of the funnel for the material stored. In other cases, stoppage is due to momentary causes, such as settling of the material, damp or frost. When the flow stops, the material is usually blocked in the bin either in the form of a bridge (an enclosed cavity not visible from the top) or in the shape of a funnel or a chimney.

The first reaction of the worker who sees that the flow has stopped is to try to restart it by pushing a pole or some other implement up through the funnel opening. This may work for a time; but then there may be another stoppage. Next, the worker goes to the top of the bin and, standing on a platform, tries to start the material moving with a shovel or a bar. If this attempt fails, the worker takes a ladder, places it inside the silo, descends, walks over the material and tries again. Suddenly, as the mass starts to move, the worker may lose footing and be engulfed, before there has even been time to cry for help. At the mouth of the funnel under the silo, the other workers will notice nothing amiss until an arm or a leg appears at the opening; but by then it is probably too late to save their mate's life.

Such accidents have happened in large silos as well as in small ones, and in silos containing sawdust or grain as well as in those containing sand or gravel. Moreover, once people are buried in a mass it is very difficult to pull them out. In one case, seven men were unable to pull out a man who was buried in sand up to his thighs; after some terrible minutes, he disappeared under the sand and died before the silo could be emptied.

To prevent such accidents, it is sometimes considered to be sufficient to provide workers who enter bins with a belt attached to a lifeline and to have another worker stationed outside to give help if necessary. These measures have proved to be necessary but insufficient if no special precautions have been taken.

Regulation 184, paragraphs 34 to 41, of the *Model code* contains the following provisions on storage bins for dry bulk material:

34. *Dry bulk material should be stored in bins which will permit removal from the bottom.*

35. *Open-top hopper bins containing bulk material which is discharged at the bottom either by hand or by mechanical means shall be covered with gratings which will allow the use of pokers to break up bridging of the stored material, but which will prevent workers from falling into the bins.*

The gratings mentioned in paragraph 35 give workers with pokers easier access to the stored materials than gangways do, because it is possible to stand on them at any point above the material. However, in very big silos, the gratings required would be impossibly large. Therefore, suitable, well guarded gangways and platforms should be installed at the top of the silo.

36. *Where it is necessary for workers to enter bins for storing dry bulk material—*

(a) *each worker shall be provided with, and shall use, a lifebelt attached to a lifeline that is as short as practicable and securely fastened to a fixed object; and*

(b) *another worker shall be stationed outside during the entire operation to render such assistance as is needed.*

Working rules should prescribe that only the person in charge of the work may decide whether it is necessary to enter a bin. In order to minimise the need for entering them, the bins should be equipped with every possible safety feature at the time of construction; for instance, the funnels should be large enough for the contents to be emptied efficiently and there should be a sufficient number of gratings at the top. When this is impracticable, openings should be made in the walls so that workers can free blocked material with poles or by other means. Depending on the circumstances, and on the kind of material stored in the bins, further precautions can be taken by equipping the bins with heating apparatus (against frost), vibrators, water jets, compressed-air jets, mechanical breakers, etc. In addition, for working on gratings or platforms, suitable rods, poles or other appliances should be readily available.

The lifebelt should preferably take the form of a harness of the kind shown in figure 36. Harnesses and lifebelts should be examined regularly to make sure that they are in good condition. The requirement that the lifeline has to be "as short as practicable" is essential. As already indicated above, it is practically impossible to pull up someone engulfed in a mass of material, and so every precaution should be taken to prevent workers from getting even their legs buried in it.

Figure 36.   The safest form of harness

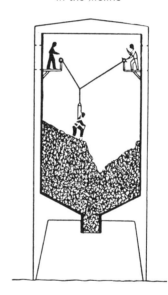

Figure 37.
There should be no slack
in the lifeline

Figure 37 shows a method of ensuring that there is no slack in the lifeline when the worker has to be in the middle of the silo (note that two attendants are required).

In other cases, the lifeline should be kept as nearly vertical as possible. As the worker changes position, the line should be let out or taken in. The attendant stationed outside should keep watch continuously, and should always be in such a position as to be able to pull up the worker inside the silo. The lifeline should be passed two or three times round a substantial fixture so that it is secure, and it should be kept "as short as practicable" while the worker fastened to the line remains in the bin. The attendant should be perfectly familiar with the hazards inherent in this type of work. The point of anchorage for the lifeline should never be changed when the worker in the silo is not in a safe position—that is, is not standing on a part of the structure or on a ladder securely fixed to it.

37. *Workers shall not be permitted to enter bins used for storing dry bulk material until all supply of material to the bin has been discontinued and precautions have been taken against accidental renewal.* As a rule, workers around a silo have no direct control of the filling equipment. Hence, special precautions have to be taken to prevent the machinery from being started unexpectedly. One such precaution is to lock the switches. It is also desirable that the discharge funnel should remain closed during work in the silo.

38. *Bins used for storing dry bulk material shall be provided with stairways or permanent ladders, and platforms where necessary, for easy and safe access to all parts, with standard railings on both stairways and platforms.* It is desirable to have fixed ladders inside bins; if portable ladders are used, they must be adequately fixed to structural parts of the bin.

On the subject of standard railings, Regulation 13 of the *Model code* (which deals with stairways) contains the following recommendations:

(1) . . . (9)

10. *All stairways having four or more risers shall be equipped with stair railings on any open side.*

(11) . . . (13)

14. *Stair railings shall be constructed in a permanent and substantial manner of wood, pipe, structural metal or other material of sufficient strength.*

15. *The height of stair railings, from the upper surface of the top rail to the surface of the tread in line with the face of the riser at the forward edge of the tread, shall be not less than 76 cm (30 in); if the railing is used as a handrail, the height shall not be more than 86 cm (34 in).*

Figure 38. A safe stairway

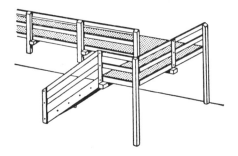

Regulation 12 contains the following provisions on standard railings for platforms:

(1) . . . (11)

12. *Standard railings shall be at least 90 cm (36 in) from the upper surface of the top rail to floor level.*

13. *Standard railings shall have posts not more than 2 m (6 ft 6 in) apart and an intermediate rail half-way between the top rail and the floor.*

In conclusion, mention should be made of the final provisions of Regulation 184, which read as follows:

39. *Bins used for storing combustible dry materials shall be of fire-resisting construction and provided with lids and an adequate ventilation system.*

40. *Where dry bulk material is piled and removed manually, undermining of piles shall not be permitted.*

41. *Special precautions shall be taken where the dry material stored is such as to lead to the formation and release of explosive or toxic mixtures.*

These last three provisions remind us that there are yet other hazards in storage bins. For instance, there is a danger of suffocation in silos where there is a shortage of oxygen; this can be expected where cereals are stored. If necessary, the atmosphere in silos should be tested before persons enter the bin, and fresh-air helmets should be worn.

ACETYLENE CYLINDERS

Acetylene cylinders are made of steel; they are filled with acetylene dissolved in acetone. They should be handled carefully to prevent damage, as this may lead to bursting of the cylinder or leakage through the cylinder valve; they should also be protected from excessive heat, which causes an increase of internal pressure and thus, perhaps, an explosion.

Figure 39. A simple carrying device for acetylene cylinders

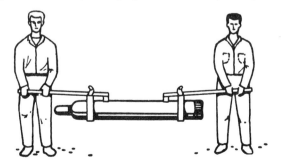

Therefore, acetylene cylinders should not be banged, jolted violently, dropped or thrown about. When being loaded on to, or unloaded from, a truck, a cylinder should be lowered gently into place. In the workroom, the cylinder should be fixed to the wall, or to a column, by a collar or a chain, in such a way that it can easily be removed in case of fire (see figure 40). This will prevent the cylinder from falling over.

Figure 40. Ways of storing acetylene cylinders

 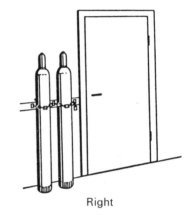

Wrong   Right

During transportation, the valve should be protected by a steel cap. When in use, the cylinder should be kept in an upright, or nearly upright, position to prevent the acetone from escaping with the acetylene.

Acetylene cylinders are normally filled to a pressure of 15 kg/cm² at 15 °C (about 200 lb per square inch at 60 °F); at this pressure, dissolved acetylene can be stored without danger. The risk of an explosion increases sharply with the pressure in the cylinder, and, consequently, if the latter is substantially above this level (say, 17 kg/cm² at 15 °C — or 230 lb per square inch at 60 °F — or higher), it is highly advisable to refuse to accept delivery and to return the cylinder at once to the filling station.

For every degree Celsius by which the temperature rises, the pressure increases by about 0.4 kg/cm² (about 5.5 lb/sq in). Therefore, cylinders should not be exposed to the direct rays of the sun or to the heat of ovens, boilers, heating apparatus, etc. In case of fire, acetylene cylinders (and other gas cylinders) should, if possible, be removed from the premises to prevent an explosion. Heating an acetylene cylinder will result in an increase of pressure not only because acetylene is liberated from the acetone, but also because of decomposition (known as dissociation) of the acetylene; it is the latter that is particularly dangerous and most liable to cause an explosion.

When, after finishing work, a worker hangs the acetylene and oxygen hoses and the torch on the acetylene cylinder, without completely extinguishing the welding flame, a small flame will remain burning near the outside of the cylinder. If this goes on for some time, dissociation of acetylene in the cylinder starts. For this reason, there should be a special support, near the acetylene and oxygen cylinders of a welding set, to hold the pipes when the apparatus is not in use.

Another cause of overheating in a cylinder may be a flame near the connection between the pressure-reducing valve and the cylinder valve, due to leaking acetylene which has been ignited by a spark or by static electricity. This may occur when the two parts have been imperfectly connected, or when the washer used for this connection is damaged; welders should always keep a good supply of new washers. Also, leakages of acetylene sometimes occur near the stem of the cylinder valve; this, however, can be prevented by the use of valves of the membrane type.

If a flashback occurs in the acetylene hose, the flame may reach the inside of the cylinder, passing through the reducing valve. Flashbacks may be caused by incorrect handling of the torch, or by a defect in the torch, such as a metal particle blocking the nozzle or a badly connected torch tip. To prevent flashbacks, a flame-arrester should be placed between the hose and the valve; but it is equally important that the equipment be used properly and kept in good condition.

Another cause of acetylene dissociation is the presence of copper in the cylinder or its accessories; this element, when in contact with acetylene, forms an explosive compound. Thus, there should be no copper in any apparatus which comes into contact with acetylene. Unfortunately, this rule is not always observed. Acetylene pressure gauges are sometimes fitted with copper Bourdon springs, or the springs are sometimes soldered with copper to the casing of the gauge. However, alloys containing less than 63 per cent of copper are not dangerous in this respect.

Thus, acetylene dissociation in an acetylene cylinder may begin either with or without external application of heat. In both cases, a part of the cylinder will become hot. What is to be done then?

If there is a small flame near the cylinder valve, it should first be extinguished and then the valve should be closed. The cause of the incident should be investigated and the defects repaired; and every five minutes for an hour the cylinder-head temperature should be checked to ascertain whether dissociation has started. If the flame is of some size, the same procedure should be followed, but in this case it will be necessary to wear gloves when shutting the valve.

If the flames are so fierce that the valve cannot be closed, they should be extinguished with a fire extinguisher, preferably one filled with carbon dioxide. If such an extinguisher is not available, a blanket or a wet cloth should be used. It is also possible — but not easy — to put out the flames with a water jet.

If a part of the acetylene cylinder is seen to be rapidly becoming hotter, the cylinder valve must be closed and kept closed. If possible, the cylinder should be thrown into a canal or a river, or be cooled by a water jet in a place where there is no special danger in case of explosion. In either case, it should be given four or five hours to cool. If it is impossible to take such measures, an explosion must be expected and the area cordoned off.

In the case of spontaneous heating, the cylinder valve must be closed and kept closed. The idea that opening the valve will prevent a dangerous increase in pressure is mistaken; it will accelerate the process of decomposition, and, furthermore, the aperture of the valve will become blocked by accumulations of soot.

The foregoing shows how essential it is to be able to close the cylinder at any moment. For this reason the key used for turning the valve spindle should always be on, or near, the cylinder.

PORTABLE LADDERS

Portable ladders are used to give access to scaffolding, platforms and other places in buildings under construction, to places where people have to go for maintenance and repair work, and for many other purposes. Ladders are, for instance, an important part of the equipment of fire brigades. In many countries, ladders are in such common use that it seems superfluous to draw attention to the purposes they serve. However, there are countries where the use of ladders is far from general and workers climb up scaffolding, using uprights, ledges and other parts of the construction; to reach a higher level for maintenance or repair work, workers sometimes use primitive and badly constructed apparatus that has hardly any resemblance to an ordinary ladder.

In countries where portable ladders are in general use, they are manufactured in specialised factories and are usually made of wood or aluminium alloy. On building sites, home-made ladders are often used. The design of a ladder depends on the use that is to be made of it. Ladders to be used on building sites have to be more rigid than ladders for maintenance or repair; ladders for window-cleaners have the rungs placed at larger intervals (about 35 cm) than is usual for ladders for general purposes (about 25 cm).

As it is desirable to keep the weight of portable ladders down to a minimum, the construction is as light as possible; hence, ladders should be handled carefully. Aluminium ladders have advantages over wooden ones, because they are extremely light. However, it is dangerous to use metal ladders near uninsulated electric wires, for all metals are good conductors of electricity.

In some countries, the construction of ladders has been standardised and the different types tested. This practice has helped considerably to ensure safe construction.[3]

Figure 41. Examples of safe ladder construction. Note how the rungs are mortised into slots in the uprights

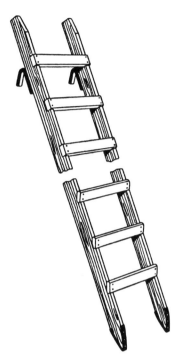

In 1937, the International Labour Conference adopted a Recommendation on safety provisions in the building industry. This Recommendation contains a model code of safety regulations which includes a number of provisions concerning ladders. For instance, Regulation 3 contains the following.

1. . . . *all ladders shall be of sound material and be of adequate strength having regard to the loads and strains to which they will be subjected.* Ladders used on building sites are usually made of ordinary wood, and consequently should have dimensions suitable for the material used and the rough treatment to which they may be exposed. An example of a good home-made construction is shown in figure 41. The dimensions of this ladder, made of ordinary pine, should be —

(a) if the length of the ladder is not more than 3 m (9 ft 9 in): uprights 5 x 7 cm (2 x 2.75 in); rungs 2 x 7 cm (0.75 x 2.75 in);

(b) if the length of the ladder is more than 3 m (9 ft 9 in), but not more than 5 m (16 ft 3 in): uprights 5 x 10 cm (2 x 4 in); rungs 2.5 x 7 cm (1 x 2.75 in).

The uprights are mortised in such a way that the load is suitably distributed between the rungs and the uprights.

The second provision reads as follows:

2. *The wooden parts used for ladders shall be of good quality, shall have long fibres, shall be in good condition, and shall not be painted or treated in a manner likely to hide defects.* Knots, and other irregularities in the fibres, may cause breakage if a relatively heavy load is placed on the ladder or if it is exposed to shocks. As weak spots (e.g. cracks) in uprights and rungs can be made invisible by painting, the painting of ladders is often forbidden. Varnishing is allowed, provided that the varnish is transparent and does not prevent inspection of the quality of the wood or its condition.

Regulation 22 of the same code contains the following provisions:

1. *Every ladder used as a means of communication shall rise at least 1 m (3ft 3in) above the highest point to be reached by any person using the ladder, or one of the uprights shall be continued to that height to serve as a handrail at the top.*

2. *Ladders shall not stand on loose bricks or other loose packing, but shall have a level and firm footing.* Having a firm footing is one of the most important safety precautions to be taken with ladders. A particular warning should be given against the extremely dangerous practice of using a ladder which is too short, and placing some object under it to make it reach higher; this has often led to accidents.

3. *Every ladder —*

(a) *shall be securely fixed so that it cannot move from its top or bottom points of rest; or*

(b) *if it cannot be secured at the top, shall be securely fastened at the base; or*

(c) *if fastening at the base is also impossible, shall have a man stationed at the foot to prevent slipping.* Not only should the danger of the foot of the ladder slipping be kept in mind, but care should be taken to prevent the top of the ladder from slipping sideways.

Figure 42.   Never do this. The ladder may slip sideways. Move it along

4. *The undue sagging of ladders shall be prevented.* Long ladders (e.g. those used for giving access to scaffolding) should be fixed by braces to the scaffolding at suitable intervals to prevent undue sagging.

5. *Ladders shall be equally and properly supported on each upright.* When positioning a ladder, take care that each of the two uprights has a firm footing, so that the load on the ladder is equally shared.

6. *Where ladders connect different floors —*

(a) *the ladders shall be staggered; and*

(b) *a protective landing with the smallest possible opening shall be provided at each floor.* It is desirable not to use ladders which are too long. Sometimes, two or three have to be used to connect different levels. If the ladders are staggered, all of them can be used simultaneously without any danger of overloading or of persons on the lower ladders being hurt by an object dropped by a person on a higher one.

7. *A ladder having a missing or defective rung shall not be used.* Refer back to Lesson 3 for an example of the kind of accident which can occur if a ladder with a missing rung is used.

8. *No ladder having any rung which depends for its support on nails, spikes or other similar fixing shall be used.* Experience has shown that rungs held in place by nails or spikes alone are unreliable. Two correct ways of fixing rungs are shown in figure 41.

9. *Wooden ladders shall be constructed with —*

(a) *uprights of adequate strength made of wood free from visible defects and having the grain of the wood running lengthwise; and*

(b) *rungs made of wood free from visible defects and mortised into the uprights, to the exclusion of any rungs fixed only by nails.* Wooden ladders made in specialised factories usually have the rungs fixed in mortises in the uprights, and kept in place by glue and by wooden safety pins which go through the upright and enter a hole about 8 mm (0.3 in) deep in the rung. Uprights should be made of wood of good quality (such as Oregon pine) and without blemishes. The two uprights of a long ladder are not parallel, but converge slightly towards the top. Rungs should be made of good-quality oak, hickory, ash, or other wood with similar properties.

10. *Roofers' and painters' ladders shall not be used by workmen in other trades.* Roofers' and painters' ladders are often of a very light construction and are not sufficiently rigid for general use.

The *Model code of safety regulations for industrial establishments* contains a general provision to the effect that all ladders should comply with the provisions laid down in the 1937 safety code for the building industry; it then goes on to recommend certain additional precautions.

Regulation 211 contains the following provisions:

(1)...(15)

16. *An adequate supply of portable ladders of good construction and of such types and lengths as may be required should be kept in readiness for use in maintenance and repair work.* Portable ladders for maintenance and repair should be of sound construction. This does not mean simply that ladders should be of suitable materials and sufficiently rigid, but also that the different rungs should be of the same construction and that the intervals between them should be uniform. To ensure that ladders are always placed in the right position (see paragraph 22 below), it is necessary to have ladders of different lengths.

17. *Ladders shall always be kept in good condition and shall be inspected at regular intervals by a competent person.* As the condition of a ladder is an important safety factor, regular inspection is necessary, especially in the case of wooden ladders, which can easily become damaged and are more likely to have hidden defects than metal ladders.

18. *Portable ladders that have missing or damaged rungs or are otherwise defective shall not be issued or accepted for use.* This paragraph covers much the same ground as paragraph 7 of Regulation 22 of the safety code for the building industry, quoted above.

19. *Defective ladders shall be promptly repaired or destroyed.* If a defective ladder is present in a workplace, it is likely to be used when somebody wants a ladder just for a moment. For this reason, defective ladders should be repaired immediately or removed.

20. *Portable ladders should be equipped with non-slip bases, when such bases will decrease the hazard of slipping.* When wooden ladders are used on concrete floors, there will usually be no risk of slipping. On the other hand, on glazed tiles, parquet floors and floors covered with linoleum, there is such a danger, and special precautions have to be taken. On these floors, non-slip rubber or lead bases should be used (see figure 43). Non-slip bases should be checked regularly, as they can become worn. Many types are satisfactory when they are new, but become quite unsuitable after being in use for some time. Most non-slip bases are useless when the floor is wet or covered with oil.

Figure 43.   Some examples of non-slip bases

21. *The person in charge of repair work for which portable ladders or platforms are required shall see that the ladders and platforms are of the proper type for the work in question.* An adequate supply of portable ladders in itself is not sufficient. It is also necessary to see that the right ladder is chosen for a particular job; for instance, when the workplace has a low ceiling, the ladder should not be so long that it cannot be placed correctly (see following paragraph).

22. *Portable ladders should be used at a pitch such that the horizontal distance from the top support to the foot of the ladder is one-quarter of the length of the ladder.* Note also the observations on paragraph 16 above.

Figure 44.   Correct pitch of a ladder

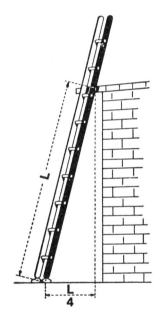

23. *Crowding on ladders shall not be allowed.* As portable ladders are made as light as possible, so that they can be moved about easily, they are not suitable for use by more than one person at a time. The practice of one person going up a ladder holding one part of a heavy object, followed by another person supporting the lower part of the object, is to be condemned; ladders are not made to stand such strains.

24. *Portable ladders shall not be placed in front of doors opening towards the ladder unless the door is blocked open, locked or guarded.* If this is not done, there is a danger that the ladder may be brought down, with the worker on it, if somebody opens the door.

25. *Portable ladders shall not be spliced together.* Splicing two ladders together may considerably increase the strain on the uprights of each, and in this way may weaken them.

26. *Portable ladders shall not be used as a guy, brace or skid or for any other purpose for which they are not intended.* It is dangerous to use ladders for purposes which may set up abnormal stresses in the uprights, since concealed defects may result which will one day cause an accident.

27. *Portable ladders shall be so stored that—*

(a) *they are easy of access;*
(b) *they can be easily and safely withdrawn for use;*
(c) *they are not exposed to the weather, excessive heat or excessive dampness;*
(d) *they are exposed to good ventilation; and*
(e) *if horizontal, they are sufficiently well supported to avoid sagging and permanent set.*

Ladders should be easily accessible to workers. Otherwise a worker may well look for something else that is not so safe (e.g. a box or a pile of objects) on which to stand, and thus be exposed to unnecessary risks.

The fibre structure of wooden ladders may be damaged by exposure to the weather, excessive heat or dampness, or by storage in a badly ventilated place. The weakening may not be immediately visible. Thus, a defective ladder, which is liable to break at any time, may be left in use.

If ladders are not properly supported when in a horizontal position, they will sag, and may become permanently deformed and weakened; of course, long ladders will be most liable to sag.

QUESTIONS

1. Mention some of the precautions that should be taken with hand tools.

2. What can be done to promote the safe use of hand tools by maintenance and repair men?

3. Why can portable electric apparatus be dangerous?

4. What are the maximum voltages considered safe for portable electric apparatus?

5. Discuss whether power-driven gears should always be completely enclosed.

6. What dangers are presented by storage bins containing dry bulk materials?

7. In what circumstances may workmen be allowed to enter storage bins?

8. Describe some precautions to be taken when using acetylene cylinders.

9. What should be done when an acetylene cylinder is found to be on fire?

10. Mention some of the requirements that a good portable ladder should satisfy.

11. What care should be taken of portable ladders to keep them in safe condition?

12. Mention some precautions that should be taken by persons using ladders.

Notes

[1] The generally accepted thresholds are: three-phase and earthed neutral alternating current (AC), 42 volts; single-phase alternating current, 24 volts; direct current (DC), 110 volts.

[2] See Ministry of Labour and National Service, Factory Department: *Accidents, how they happen and how to prevent them,* Vol. 11 (New Series), Apr. 1952 (London, HMSO).

[3] See, for instance, British Standards Institution: *Aluminium ladders, steps and trestles,* British Standard 2037 (London, 1964, amended 1981).

PSYCHOLOGICAL AND PHYSIOLOGICAL ASPECTS OF ACCIDENT PREVENTION 8

The last three lessons were concerned with some of the technical aspects of accident prevention. We discussed the workplace and its equipment rather than the role of the worker. But, as we have already seen in Lesson 3, one of the main factors in the chain of events that leads up to an accident is the worker. A large number of accidents over the years have been blamed on unsafe acts performed by workers — carelessness, fatigue, boredom and inattention are just some of the suggested reasons for accidents. How often has the carelessness of the worker been put forward as the cause of an accident — "It was my fault, I was careless"? This approach has done little to focus on the root causes of accidents. It suggests that all that has to be done to prevent accidents from occurring is to stop unsafe acts. Thus, the onus is often incorrectly put on the worker, and the conditions that have resulted in the unsafe act are not given full consideration.

In any discussion of the human factor, it is usual to talk about physiology and psychology but, as is so often the case, no one is sure where the one begins and the other ends. It might be more accurate to treat mind and body as a single entity, and humans as psycho-physical beings. The subject is infinitely complex, and there is room to mention only a few of the numerous elements in "the human factor" which operate to endanger or to protect the worker.

In this lesson, therefore, we will deal only with general matters, and with some problems which are frequently discussed, such as industrial relations, accident proneness of individuals and fatigue. Some means of prevention that are considered to be in the psychological domain — propaganda, education and training — will be discussed in Lesson 9.

ATTITUDES TOWARDS SAFETY

There are many possible answers to the question — why do workers do a job in an unsafe way when they could do it in a safe way? The workers may consider the unsafe way easier, less troublesome or faster; they may think that the unsafe way is the best one; they may consider safety precautions unnecessary because they are convinced that they can look after themselves in all circumstances; they may feel that, as people of experience, they can quite well determine their own way of working; or they may be ignorant or unaware of the safe method.

It may be assumed that, as a rule, a factory worker desires to earn the highest wages possible. If wages are higher when work goes faster, it is not difficult to understand that the worker may be led to neglect safety in order to increase output. It is also easy to understand that many people will not give up an easier way of working (i.e. one that gives the same results with less trouble) just because another way is safer. These considerations have already been mentioned in the discussion in Lesson 5 on the technical aspects of accident prevention. If it is more comfortable to work without a guard (as is sometimes the case with certain old-fashioned guards on woodworking machines), or if a better finish can be given to the job if the guard is removed, a worker who is highly skilled, and who takes pride in his or her work, may well object to working with it.

If workers are in a factory with many other persons, under supervisors, overseers and managers, they may feel that, while it is necessary to obey orders, they do not want to do so more than is necessary. They may want to control their own workplace and job, and to take care of themselves. They may object to interference in personal matters, even to orders given in their own interest. This attitude, which is often mere bravado, can, however, be dangerous and lead them to take risks just to prove their independence; it should have been counteracted when they were at school or learning their trade. They also should have been taught safe working methods. Many of the difficulties in accident prevention are due to inadequate vocational training.

When considering why so many persons do not co-operate in ensuring maximum safety, we should not forget another point. A person needs a powerful imagination to realise what an accident can really mean; it is very difficult, if not impossible, to realise what losing a leg, or losing one's eyesight, means, and to put oneself in the place of a person who has.

Let us examine some of these aspects of behaviour in everyday terms, and consider how, and why, they contribute to unsafe acts.

1. *Time and safety.* One of the most common reasons for taking risks at work is to save time — time for more leisure, time to enable one to earn more money, or simply time saved by rushing to finish the job.

It is not surprising, therefore, that this wish to save time often results in an unsafe act.

2.   *Effort and safety* — or more simply "taking the easy way out". If the safe way of doing a job demands too much effort, whether it be physical or mental, it is not unnatural that workers should take short cuts; again, this may result in an unsafe act.

3.   *Group acceptance and safety.* How often has a new worker joined a group of experienced workers and, on asking about some article of safety clothing, for example, been told — "You don't need that; none of us wear them". The new worker, not wishing to be an outcast, usually ignores his or her fears — sometimes with dire consequences, because it is often new employees who are most at risk.

These few examples illustrate the everyday nature of accidents and show how easily they could happen to us. There are few of us who have not, at some stage of our working lives, cut corners to save time and effort. If we have not had an accident, we are indeed fortunate.

The general psychological patterns of individuals are not exactly the same in all parts of the world, and, consequently, the attitudes of workers to safety problems differ somewhat from region to region. In countries in the early stages of industrialisation, workers are often unaware of the possibilities of accident prevention, and consider accidents more or less as they consider diseases — namely, as unavoidable, mysterious afflictions that have to be suffered like bad weather. This attitude can be greatly altered by emphasising safety during education and training courses.

ENVIRONMENT AND FREQUENCY OF ACCIDENTS

The worker's environment is an important psychological element in promoting safety. It has been observed in many undertakings that the frequency of accidents is influenced by the general atmosphere. When relations between employers and workers are bad, when workers are dissatisfied with wages, working hours or other labour conditions, the number of accidents tends to increase, while in periods when industrial relations are good the opposite seems to be the case. Welfare measures designed to make the worker's life more comfortable also make it safer.

Workers' behaviour usually reflects their material and psychological environment. Reasonable wages, good human relations inside the factory, a good relationship between management and labour, correct decisions on questions of promotion, and, at the same time, well cared for workplaces, sanitary facilities which are much better than "good enough for a worker", and welfare facilities — all these features, both material and intangible, influence a worker's behaviour and have been found to be conducive to greater safety. As the same factors have a considerable influence on labour turnover, they also increase safety indirectly by bringing about a greater stability in the workforce.

Insecurity of employment is almost certainly a cause of accidents. When workers fear dismissal, they may well be in an emotionally unbalanced state, which will make them more liable to have accidents. In countries with social security legislation which guarantees payment of at least a part of wages in cases of absence from work caused by an accident, but not in cases of ordinary unemployment, accidents are even caused wilfully to provide an income during an impending period of unemployment. However, such cases are not within the scope of ordinary accident prevention work.

Good order, good housekeeping, and adequately guarded machines are examples of environmental factors that not only contribute materially to safety but also have a considerable psychological effect. People in a dirty, untidy workroom, cluttered with tools, materials and rubbish, tend to behave differently from when they are in a clean, attractive workroom where good housekeeping is the rule. This difference in behaviour will usually be reflected by a difference in accident frequency.

Respect for workers' feelings and dignity helps to give them peace of mind, and this is a most important psychological safety factor. The workers are more likely to be immune from injury when the management is careful in its relations with them as people.

However, peace of mind does not depend only on the situation inside the factory. Living conditions outside the factory also have an influence. Living in slum areas, for example, has undesirable moral and physical effects, and thereby adversely influences the worker's attitude in the factory; hence, accident rates tend to increase. Here, again, it is clear that safety is not something that can be separated from other aspects of life; it is linked to them all.

The worker's peace of mind may also be disturbed by family circumstances. Remedying these situations is a very delicate task. Of course, it is most undesirable to do anything that might give the impression of interfering with a worker's private life. Something can, however, be done if there are social workers in the personnel department of the undertaking.

FATIGUE AND BOREDOM

Many authorities consider that fatigue increases the risk of accidents, and, the greater the fatigue, the greater the risk; but the relation between fatigue and accidents is extremely complex, and it is not easy to draw simple conclusions.

One of the most common complaints among workers is tiredness, whether it be due to lack of sleep the night before, overwork or some emotional problem. When the tiredness becomes so pronounced and persistent that it interferes with work, and with activities at home, it is termed "fatigue". Exhaustion is an extreme state of fatigue. Many shiftworkers suffer from fatigue, largely as a result of disturbances in their normal 24-hour cycles, or biorhythms. Most of our normal mental and physical processes are geared to the time of the day or night and to a specific part of this cycle. Thus, if we have to stay awake at night, we tend to feel tired and lethargic because our body expects to rest at that time and not to undertake physical or mental work. We find it difficult to adapt to such changes in our daily lives.

It is not surprising, therefore, that many studies have shown that workers tend to make more mistakes, and to work at a slower rate, on the night shift. Not all studies, however, have confirmed an increase in the rate of industrial accidents on the night shift. What has been found is that, whereas all accidents tend to be more frequent during the morning shift, there are more severe accidents on the night shift. Also, as minor accidents on the night shift are often not reported, the accident rates are misleadingly low.

In addition, the influence of fatigue differs with different persons. Workers who are very interested in their jobs will give all their attention to them and will not feel fatigue, while workers who are not interested at all in their jobs tend to become inattentive and careless at times when they cannot be suffering from fatigue.

Accidents are often related not so much to physical fatigue as to the mental attitude of the workers. This attitude depends to a high degree on whether they like their work or not. Here, again, the general environment counts. Everything which helps to augment their interest and satisfaction in their work — such as responsibility for the work, being given appreciation by the management, and being kept informed of what goes on in the factory — will tend to decrease their liability of having accidents. It is well known that periods of boredom can be tolerated more easily in an enjoyable social setting — a point that is seldom encouraged by managements.

There are some workers who may be quite satisfied with monotonous work; they do it almost automatically and without thinking. Accidents sometimes happen to them when, for one reason or another, something goes wrong with their machines or their workpieces, while they continue the usual movements automatically. This has happened many times to workers on metal presses. If there is a defect in the clutch mechanism of the press, the ram may descend a little later than usual; the worker's hand is already under the die to remove the workpiece and is hurt by the descending ram. In most cases, such accidents can be prevented by the installation of a suitable guard.

Other workers find monotonous work intolerable; they cause accidents by trying to find variations in the ever-recurring cycle of movements. In such cases, accidents occur because the workers are too intellectual for the type of work they are doing, and not because they are "playing with the machine", as the findings of accident investigations sometimes state.

EXPERIENCE AND INEXPERIENCE

It is equally difficult to draw clear conclusions about the influence of the length of service and the experience of the worker on accident rates, for the different factors which lead to accidents influence one another and cannot be isolated. For instance, the majority of inexperienced workers are adolescent and the majority of experienced workers are adult; as mentioned in Lesson 3, it may be difficult to distinguish between age and experience.

The attention of workers who are not yet familiar with the factory environment will be distracted by many new impressions, and this, together with their lack of experience of the job, may explain the relative frequency of accidents among newcomers. Much depends on the education and training they receive before entering industry, and on the way in which they are introduced to their jobs and supervised.

On the other hand, although experienced workers are not handicapped by unfamiliarity with their surroundings, their very familiarity with the risks of the job often makes them less careful. A characteristic example is the high number of electrical accidents occurring to electricians. Moreover, if no serious accidents occur in a particular type of work for a considerable time, the workers become less careful, for they gradually tend to assume that the danger is not as serious as they have been told. Safety measures will then be neglected until another accident shows the importance of safety precautions.

ACCIDENT PRONENESS

As we said earlier, one of the most popular concepts in accident causation used to be "accident proneness". This theory was based on observations that, whereas some workers had no accidents, others had several in a given period of time; in other words, some workers were more liable to have accidents than others—hence the term "accident prone". Although this theory received considerable support at one time, it has now largely been disproved. Rather than being accident prone, it is far more likely that some workers are victims of the law of probability.

PHYSIOLOGICAL CONDITIONS

Some accidents are attributable to the worker's physical condition. Many people have eye defects without knowing it; others suffer from a certain degree of deafness; still others suffer from illnesses (e.g. epilepsy) which expose them, and other people, to abnormal risks. Factory doctors should give special attention to such cases. If the handicaps cannot be eliminated, suitable work should be selected for those suffering from them, if possible. The problem is not how to exclude such persons from work, but how to employ them usefully despite their physical defects or infirmities. If a doctor is not available, the management should try to find a solution. It should be remembered that accidents may be caused by physical defects.

QUESTIONS

1. Give some of the reasons that might lead workers to do dangerous things.

2. Discuss whether workers who are satisfied with their conditions of employment are less liable to have accidents than workers who are dissatisfied with them.

3. Can fatigue lead to accidents? If so, how?

4. Are experienced workers more careless than inexperienced ones?

5. What do you know about accident proneness?

6. Is there a relation between health and accident proneness?

PROPAGANDA, EDUCATION AND TRAINING

9

This lesson is concerned with some educational measures which are aimed directly at the worker, and are applicable to all workers, irrespective of their physical or mental make-up.

Three types of educational measures have been used over the years: propaganda, education and training.

Propaganda includes stickers and posters, film shows, talks, competitions with rewards, safety weeks, and so on. These will be described in turn. Although many would consider that safety propaganda is a rather outmoded means of drawing attention to the real causes of accidents, or of promoting safety, and that it has only a limited value in stimulating and encouraging people to work safely, it is still widely practised in many countries. In some countries, developing countries in particular, it may well be the only means of promoting safety.

As the attitude that accidents must be prevented before they occur has gained ground, people have come to realise that education and training are vital in the promotion of safety. Such education and training may begin in primary school, and may be continued in secondary schools, colleges, education centres or on site, according to the courses and the target groups.

Practical training can also be given in various places: in trade and technical schools, in apprentices' courses and on the job.

Propaganda seeks to persuade, education seeks to provide information, and training seeks to provide skills, but there is really no sharp distinction between them, and all have some educational value.

PROPAGANDA

Posters

There are all sorts of safety posters and each may help to promote safety in a different way. Some are humorous, some gruesome, some give general advice, some demonstrate a particular hazard in a particular operation, and so on. Posters may be used to discourage bad habits which are widely practised, to show the general advantages of working safely, or to give detailed information, advice or instructions on particular points. Some try to influence workers by appealing to their pride, self-love, affection, curiosity or humour.

One type of poster (known as the positive poster) shows the advantages of caution; another type (known as the negative poster) illustrates the consequences of carelessness. Those in favour of positive posters think that, if workers are shown a good example, they will follow it. A positive poster may be of value, for example, at a time when workers are worrying about something—perhaps family troubles—and their attention wanders from their work. At such times they need encouragement, and it is therefore inadvisable to use negative posters which may cause fear or, as it is sometimes called, "safety neurosis". Those who encourage the use of negative posters, on the other hand, think that, as workers do not realise the hazards to which they are exposed, they should be shown the risks in a realistic way. They also believe that, to achieve results, deeper impressions must be made than can be made by positive posters.

Figure 45 shows an example of a negative poster, published by the Egyptian Ministry of Commerce and Industry, and figure 46 illustrates the positive approach.

Figure 45. When carrying iron bars or rods, beware of overhead power lines

Figure 46. Condition in which hammers should be maintained

Both groups agree that safety posters should be chosen according to the workers' capabilities, which will obviously differ from industry to industry and from country to country.

Experience has shown that it is very difficult to design safety posters which have more than a momentary effect on workers. Several proposals have been made to overcome this difficulty. One suggestion is that the artist who designs a poster should make sure that the details are right, a technician should verify that the poster is technically correct, and a psychologist should advise on whether the desired impression will be made.

Some people recommend photographic posters, others prefer drawings. Photographs show real events and in particular provide a means of illustrating situations in the undertaking concerned. Drawings have the advantage of showing the one thing that needs special attention and omitting all non-essential details. Figure 47 is an example of a good, simple drawing, used on a poster published in the United Kingdom.

Safety posters should be displayed in places where workers usually spend some of their time when they are not working, such as the factory entrance and locker rooms. People disagree over whether posters should be displayed in canteens. Some people think that canteens should be places for rest and recreation, in complete contrast with the factory itself, and that posters reminding workers of the workshop, and of their jobs, should not be displayed there.

113

Bulletin boards for posters should be agreeable to the eye and properly maintained. They should be attractively painted, glass-fronted and well lighted. To stimulate the interest of the workers, notices and objects relating to safety in general, or to a recent accident, should be displayed on the same board. Such objects may be, for instance, safety shoes or glasses damaged during normal work. Only a small number of different posters should be displayed simultaneously, and they should be changed periodically, at intervals of about one or two weeks; even daily renewal has been proposed.

Safety posters can only be an accessory to improving safety; they cannot replace good housekeeping, correct planning, good working habits and suitable guards, but they should help to create a greater awareness of safety among workers. One advantage is that they can be used anywhere, both in large undertakings and in small workshops, at little cost. In many countries, this may be the only method employed to promote safety.

Posters can be useful for instruction. If, for instance, a special type of guard has not been adjusted correctly — as is often the case with the riving knife of a circular saw — or if an unsafe working method is being followed (e.g. if loads are slung on the hook of a crane in the wrong way), the workers should be shown the correct working method; but, at the same time, posters can be used as a reminder of the correct way of dealing with these matters. Below are two typical examples.

Figure 48. Never take a chance with worn ladders

Figure 49. The assistant should wear protective clothing too

Safety posters should not be used to show unsafe conditions when such conditions are the responsibility of the management. For example, there should be no need for a poster warning workers about unsafe power presses, since the management is responsible for installing guards on the presses and for keeping them in good working order. The poster in figure 49 does not show whether the welder's assistant has been given adequate protective equipment for the eyes. Posters of this kind have real value only if they are designed to urge workers to use protective equipment that is actually available.

Management should never use hazard-warning posters with the intention of later denying responsibility for accidents because the warning was not observed.

Films and slides

A poster gives just one impression of a hazard. A film can tell the whole story of an accident, showing the environment, how the dangerous situation arose, how the accident happened, what the consequences were, and how it could have been prevented. As many people like watching television and going to the cinema, accident prevention programme organisers have tried to exploit the possibilities of films to make people safety conscious. Here, problems similar to those already mentioned in connection with safety posters arise, and the same means of solving them have been tried. Humour is often introduced into a film, or a poster, to counteract the viewers' reluctance to receive orders and advice.

The situation in a factory should be shown accurately to prevent the film from giving the impression that it is not based on normal working conditions. The feelings of the workers, and their customs and circumstances, should be faithfully reflected. This is particularly important, because however good a film is, it will be counter-productive if the workforce cannot relate to it. For example, if films are sent from the developed to the developing countries, improvements in safety will not necessarily follow. The film is likely to have little impact because the viewers will not be able to associate with it. The conditions will be different, the workers may have a different coloured skin, and so on. The same film is not necessarily suitable for different countries.

Films made specifically for instruction are more valuable than those made for general propaganda; they are particularly useful for explaining new safety devices or new working methods. Films can give explanations, show laboratory tests, analyse technical processes, explain difficult and complicated matters in a methodical way, and reproduce a rapid sequence of events in slow motion. Nevertheless, demonstrations will often make a more vivid impression, and will allow the audience to ask questions and discuss particular points.

Slides have some advantages over films: they can be projected for as long as required, more detailed explanations can be given, and questions can be asked. However, slides have the same limitations as posters.

Sometimes, it is a good idea to combine films and slides. A film which gives a general impression of the subject may be followed by slides which bring out the main points of the film; these points can then be discussed in detail.

Television can also be used for industrial safety education, for scenes in the factory itself can be shown in the lecture room and the audience can discuss questions arising from it without disturbing work in the factory.

Talks, lectures and conferences

With talks, lectures and conferences, much depends on the speaker's understanding of the audience. If the speaker knows how to hold the attention of those in the audience, he or she may have some influence on them. What is most essential is that the audience should feel that the speaker is sincere. For instance, little can be expected from a speech by the manager of an undertaking if it has been written by the safety engineer and if the manager does not seem to be very familiar with its contents or seems to be in a hurry to finish it in order to do something else.

As with safety posters, films and other means of safety propaganda, talks and lectures and conferences can only contribute in a modest way to safety, but they do offer an opportunity for direct contact between speaker and audience, and this is a great advantage.

Studies have been made of the value of talks and lectures, and the results have not been encouraging. Increasing use is now being made of discussion groups, in which subjects are discussed either by all present or by a panel of persons who bring out the different aspects of the subject for the benefit of the audience.

Competitions

Many people get great pleasure from sports competitions; the idea of safety competitions is obviously one which must have considerable appeal to organisers of safety programmes.

Competitions are usually organised between factories working under similar conditions or between different departments of the same factory. The department showing the best results is usually awarded a prize (for instance, a cup), which remains in that department until the end of the next competition period, when it goes to the new winner. The success of a competition does not depend on who is the winner, but on the decrease in accident rates in the factory as a whole.

Safety competitions are often held in some countries, but they are practically unknown in others. In countries where they are common, it is believed that they make a useful contribution to safety; but they lose much of their value if competitors cheat by not reporting injuries.

Exhibitions

Exhibitions are a means of acquainting workers in a very realistic way with hazards and means of eliminating them.

For example, there are a number of permanent safety exhibitions in safety museums. However, their value is limited. Because of expense, it is impossible to keep them up to date. Only a small number of those people who have anything to do with industrial safety — and, above all, very few workers — visit them. And it is difficult to prove that working systems demonstrated in a museum will be effective under normal working conditions in factories and workshops.

One method of publicising an exhibition is to invite employers and workers to visit it; but a much more effective method is to take the exhibition to them. Examples of this approach are itinerant and factory exhibitions. These usually deal with a very limited number of subjects, and are organised either by safety associations, acting on behalf of local industry or a single undertaking, or by an undertaking itself. The best results can be expected when such an exhibition is combined with other safety activities which have a limited aim. For instance, the management of a factory in which large numbers of accidents causing eye injuries occur may decide to organise an exhibition of appliances for eye protection as part of a campaign against such accidents. In another case, it may be of interest to organise an exhibition which shows how scaffolding parts should be assembled and used.

Attention can also be focused on safety questions by an exhibition of objects dealing with recent accidents in the factory — for example, a broken grinding wheel, the flying pieces of which were caught by the hood, or a hard hat damaged by a falling object (the object should also be shown). Such an exhibition shows the practical value of safety precautions. It is most important that the means of protection exhibited should be available to the workers. Otherwise, the workers will, with some justification, wonder why such protective equipment is not available, and they may well feel that they have a legitimate grievance.

Safety publications

The types of safety propaganda so far described can be used for illiterate, as well as literate, workers. Documents, unless they consist entirely of pictures, are only of value to workers who can read.

There are a great number of safety publications, and they deal with a wide range of subjects. In many countries, safety magazines appear regularly and contain illustrated articles describing new safeguards, the results of investigations and research in the field of industrial safety, new ways of preventing accidents, and so forth. There are other periodicals designed not so much to pass on new knowledge as to disseminate knowledge of existing techniques to a wider public. It is important that editors of periodicals should be aware of their responsibility; in particular, they should make sure that what is recommended is really based on experience and practice. All too often articles describe safety measures which have been used only in a single case, or which are not in actual use at all, but have merely been proposed. Such articles may do more harm than good. The reader, of course, before taking any steps to make use of the ideas contained in an article, must judge whether they are applicable in his or her particular context, and whether enough experience is quoted to suggest that the ideas would have a reasonable chance of being successful if they were introduced.

Some space is usually reserved in works magazines for safety subjects. The articles on safety are sometimes intended to be read not only by the worker but also by the members of his or her family. Some people think that the family will exert an influence on the worker. The idea of bringing safety into the worker's home also underlies radio talks and television shows on safety, and the distribution of safety calendars.

Other forms of safety propaganda include pamphlets and leaflets, safety stamps, illustrations and slogans on pay envelopes, and so on. In so far as this material brings new ideas to the workers' notice, it has a certain value. If, however, such propaganda only reminds workers of what is already common knowledge (with the intention, perhaps, of dispelling indifference), it is difficult to measure its value.

It must not be thought that documentation on safety is confined to magazines, pamphlets, leaflets, etc., designed primarily for workers and their families. There is a very large, and steadily increasing, quantity of printed material, consisting of reports of labour inspectorates and research institutions, general safety manuals, manuals on particular subjects (e.g. electricity, boilers and fire protection) and all kinds of technical booklets, pamphlets, data sheets, and so on. The ILO itself has published a considerable number of codes of practice and guides to safety and health in various industries; new volumes appear regularly in its long-established Occupational Safety and Health Series, and its *Encyclopaedia of occupational health and safety* is a classic work in its field. Details of some of these publications are given in Lesson 13 and at the end of this manual. Scientific and technical material of this kind is useful to inspectors, managers, safety engineers, safety associations and, indeed, to everyone bearing some responsibility for the promotion of occupational safety.

Safety drives

From time to time, it might be necessary to launch an intensive safety drive. One way of doing this is to organise a safety day or a safety week, on a national scale, in a city, or perhaps in just one undertaking. When a campaign is organised for a number of industrial undertakings, the programme is usually a general one; but when it is organised for one undertaking, it can be focused on one particular subject. It may well make use of a combination of some of the different items of safety propaganda discussed earlier in this lesson.

National or regional safety days and weeks are usually patronised by the public authorities (such as the Ministry of Labour, or the governor of a state or province) and are given much publicity by the press, the radio, the cinemas, etc. The campaigns may include exhibitions, film shows, demonstrations, competitions, discussions, and so on.

A factory programme may be based on a single theme, such as safety shoes, lifebelts or other items of safety equipment. The programme of a safety day designed to introduce some new equipment might include an exhibition of the appliances, talks and films, to impress on the staff the importance of the measures to be taken; there might be competitions and crossword puzzles, which mention the theme in some way or another; and there might be a programme of entertainment for the workers and their families. At a final meeting, everyone will be asked to co-operate in preventing accidents of the type to which the theme of the safety day is related. On the next day, the new safety equipment will be distributed.

It is important, at this stage, to remind ourselves that, whatever the type of propaganda used, many people feel that it achieves very little in terms of safety. It does little to bring about constructive changes which will eliminate unsafe conditions. The root causes of accidents are rarely recognised and controlled by this method. Yet, in many countries, this type of propaganda is sometimes the only method employed to create safety awareness.

WORKERS' EDUCATION AND TRAINING

Since the early days of the Industrial Revolution, the primary concern of trade unions has been collective bargaining over the issues of wages and hours of work. However, the daily toll of accidents and industrial disease on their members has inevitably brought the subjects of occupational health and safety into the sphere of union activity. Initially, this involvement took the form of securing compensation for those workers who were crippled or maimed through their work. But gradually a more positive, long-term approach has been adopted, and unions have

119

started occupational health and safety education programmes for workers and their representatives. At the ILO Meeting of Consultants on Workers' Education in December 1979, the following was reported:

An obvious priority requirement is the education of elected and potential trade union representatives – shop stewards, other spokesmen of the workers and works councillors. This education must also include matters concerning the working environment and occupational health and safety.

Workers' education programmes are not new. They have been developed over a number of years in a number of countries to assist trade unions and workers' education bodies to develop their own education activities. Such programmes allow workers and their representatives to improve their knowledge on a whole range of social, economic and labour issues, thus enabling them to fulfil the functions of responsible trade unionists. Workers' education is also essential if workers and their representatives are to play an increasingly effective role in the social and economic development of their countries. The programmes vary from country to country, and from year to year, depending on the needs of the specific target groups. The people who attend workers' educational programmes may be illiterate, or they may be highly trained technicians – the whole spectrum is covered. At the time of writing, one of the most important target groups is workers' educators. Whatever the target group, the common theme of workers' education programmes is to help workers to help themselves to become better trade unionists, and so to contribute to the development of society as a whole.

Accordingly, there are workers' education programmes in occupational health and safety in many industrialised countries. Workers and their representatives are given a basic grounding in the subject so that they can participate effectively in finding their own solutions in their own workplaces. However, it is important to stress that, although knowledge may be a means to power, it is a poor lever for change on its own. Workers and their representatives must be able to use their knowledge for effective action, and to participate fully in the decisions that affect their own health and safety. Obviously such participation demands a strong and active trade union organisation, both at local and national levels.

The question now arises as to whether education in occupational health and safety should concentrate solely on the needs of workers in specific industries, or whether it should present a more general view that enables workers and their representatives to participate effectively in social, political and economic matters as a whole. The answer lies somewhere in the middle. Of course, any workers' education programme must include information on the dangers of injury and disease in a particular working environment.

Training should also be given in the laws relating to health and safety in particular industries, in approved methods of prevention, in how to carry out safety inspections, and so on. For example, in the United

Kingdom, the ten-day safety representatives' courses organised by the Trades Union Congress include seven main themes:

- the safety representative and union organisation;
- health and safety and the law;
- information for safety representatives;
- skills for safety representatives;
- work hazards, e.g. fire, lighting and lifting;
- noise at work;
- chemical hazards at work.

Several other countries (for example, Brazil, Canada, France, the Federal Republic of Germany, Norway, Sweden, the Union of Soviet Socialist Republics, the United States and Yugoslavia) have all developed similar education and training programmes specifically for workers.

In developing any such workers' education programmes in occupational health and safety, it is important that workers' organisations know where to seek advice, for this is a rapidly changing field of study. After all, no single workers' organisation can be expected to have all the specialised knowledge and expertise required to begin workers' education programmes in occupational health and safety.

When we consider the actual composition of workers' education programmes, it is important to remember that health and safety do not stop "at the factory gates" as many would believe. It is very short-sighted to concentrate solely on the improvement of the working environment to the detriment of the living environment and social well-being. Campaigns to "clean up" certain industries have in many cases simply transferred the problem to the neighbouring society.

As to the methods that should be employed in such education programmes, this depends to a large extent on the target group and on the subject. It is not within the scope of this manual to discuss the various methods that should or should not be used. However, the reader is advised to look at the following two ILO publications: *Workers' education and its techniques* (Geneva, 1976); and *How to improve workers' education* (Geneva, 1976).

QUESTIONS

1. Describe some of the different forms of safety propaganda.

2. How can safety posters contribute to accident prevention?

3. Compare the educational value of posters, films and slides.

4. Do workers' education programmes in occupational health and safety exist in your country?

SPECIAL CATEGORIES OF WORKERS

10

So far we have been concerned with aspects of safety and accident prevention that apply to all workers; but there are certain categories of workers which need special consideration because, for one reason or another, they are exposed to special risks. In many countries, some of these groups—the young, the elderly, the handicapped, and women workers—are the subject of special legislation; but in others no such legislation exists, or, if it does, it is rarely enforced. Let us examine two of these groups of workers in more detail.

YOUNG WORKERS

It is estimated that young people under the age of 20 make up about one-third of the population in the industrialised countries and one-half in the developing countries. Many of these young people are unemployed and untrained, particularly in the developing countries, and, as we have already seen, even if they are employed, the level of safety training may be inadequate.

Young workers need special care for both physiological and psychological reasons. They usually lack the physical strength of the adult, and they lack experience.

The ILO has always been concerned with the problems of young workers. It is still common to find child labour in a number of countries, in spite of a continuous effort by the ILO, which has included the setting of standards regulating the minimum age for admission to work and the recruitment of young workers into unhealthy or dangerous jobs. Two international labour Conventions, adopted as long ago as 1937, concerned the minimum age for admission to industrial and non-industrial employment, respectively, and prohibited the employment of children under the age of 15 years in industrial and non-industrial work, with the provision that different minimum ages could be fixed in certain circumstances. Other Conventions on the subject have subsequently been adopted, culminating in the Minimum Age Convention (No. 138) and Recommendation (No. 146) which were adopted in 1973. Unfortunately, Convention No. 138 has so far been ratified by only a handful of countries, despite a strenuous effort by the Organisation to obtain maximum ratification. The minimum age ranges from 14 to 16 years, or even higher, in some of the industrially developed countries, to 12 or 13 years, or even lower, in those countries in which industrialisation is just beginning. In the latter countries, minimum age laws are often inadequate or not enforced because of a shortage of schools, family poverty, lack of properly equipped inspection services, and so on. In some such countries, children as young as 5 years of age may be allowed to work in industrial undertakings, exposed to all the risks of industrial work, and fatal accidents among them are frequent. The prohibition of the employment of children under a given minimum age, realistically fixed in the light of national circumstances, is a vital necessity for the protection of their health and safety.

In several countries, the law restricts young people from working "in dangerous conditions". This phrase covers work that directly exposes workers to danger (e.g. work that has to be done at considerable heights on buildings under construction); work that brings the worker into contact with very toxic substances, such as white lead or cyanides; work classified as "heavy", such as stoking steam boilers; work that lays on the worker great responsibility towards fellow workers, such as the operation of certain machines; work for which adequate protection does not yet exist, such as certain types of work on metal presses; and work that demands a particularly high degree of concentration, such as work on electrical installations. Restrictions are also frequently placed on the employment of young people in some types of work which could clearly be harmful to their health.

A good many of the laws and regulations specifying the dangerous, heavy or unhealthy jobs prohibited to young persons include exceptions designed to enable a young person to obtain suitable training for such work. Most of them provide, too, for the lists of prohibited jobs to be revised from time to time, as technological progress may reduce the danger of accidents occurring to young people.

Young people may be protected by the managements of undertakings, as well as by statutory prohibitions, and they can follow a number of occupations, provided that safe working conditions are maintained. Safe working conditions imply correct working habits, and great care should be taken to train young people in such habits during the period of apprenticeship.

Safety can never be considered as an accessory to a working method; it should be incorporated in the method itself. This rule should be followed not only in technical schools but also in training courses for newcomers inside the factory, and whenever the apprentice has to learn his or her work by helping an experienced worker. If we consider the elimination of child labour as the first step towards greater safety, the second is to provide young people with adequate safety education and training.

Finally, when young people leave school or finish their training courses and enter the factory to start work, the safe working methods they have been taught are all too often "looked down on" by the older workers; this attitude often intimidates the new arrivals, and, in order to "be like the others", they abandon their safe working methods and follow the general practice. Much safety education has been wasted because of such attitudes. The teaching of safe working practices is not sufficient; conditions in the factory itself must encourage their use, and they should be practised until they become second nature. The question of safety and group acceptance was discussed in an earlier lesson. Attention is drawn to *Occupational health problems of young workers* (Geneva, 1973), a publication in the ILO Occupational Safety and Health Series, and to *Young people in their working environment* (Geneva, 1977).

WOMEN

Women working outside their home contribute more than one-third of the world's total labour force. Everywhere, they are faced with the problems of discrimination in the labour market. In many countries, women are excluded from certain types of employment, not on health and safety grounds, but for reasons of discrimination. There is an increasing trend for women, who were traditionally employed in agriculture and handicrafts, to work in factories.

Generally speaking, the safety rules are as valid for female workers as they are for male workers. However, special additional measures have to be taken to protect women from certain occupational risks associated with pregnancy.

An indirect accident risk arises if women employed in factories are permitted to take their children to work with them (as may happen in developing countries), because children may play near running machines or come into contact with dangerous substances. This used to be the case in some cotton-spinning mills when women working on winding machines took care of their children at the same time; some of the children played on the floor between the machines during the whole working day. In plants for sorting uncleaned wool, women in charge of sorting used to have their babies with them, and somewhat older children helped them in their work. The children were exposed to a very real risk of anthrax.

In all such cases, a crèche would not only provide proper care for the children but would also be a means of preserving them from accidents.

Accident prevention

Over the years, the International Labour Conference has adopted a number of Conventions and Recommendations dealing with the occupational safety, health and welfare of women workers. These range from the White Lead (Painting) Convention, 1921 (No. 13), to the Benzene Convention, 1971 (No. 136), and many more besides. Full details are given in an ILO publication entitled *Standards and policy statements of special interest to women workers* (Geneva, 1980).

QUESTIONS

1. What special measures, if any, should be taken to protect young workers from having accidents, and why?

2. What preventive measures are called for by the employment of women?

SAFETY ACTIVITIES
IN THE UNDERTAKING

11

Laws, regulations, inspection, recommendations, advice, research, exhibitions, congresses, and so on, will not serve any useful purpose if, in the last resort, nothing is done to promote safety in the factory itself. As we have already seen, statistics of occupational accidents have more practical value at the plant level than at the national level. It is within the factory itself that accidents must be prevented.

THE ROLE OF MANAGEMENT

It cannot be repeated too often that safety starts from the top of the organisation and works down. Everybody in the undertaking should know that the employer is interested not only in production, in the quality and quantity of the products, in preventing wastage of material and in the proper maintenance of machines and tools, but also in safety. It is essential that management should view any safety programme as part of the overall plan for the company and that they should treat it in just the same way as they would a production programme or a quality-control programme. The costs of such programmes must be taken into account and should appear on the balance sheet. Management has to organise processes efficiently, combining a maximum of production with a minimum of cost, and should treat safety not as an extra but as part of the process itself. Management has a responsibility to ensure that unsafe conditions and unsafe acts do not occur.

The following paragraphs, taken from the Robens Report which led to the Health and Safety at Work Act, 1974, in the United Kingdom, clearly identify the role of management:

46. Promotion of safety and health at work is an essential function of good management. We are not talking here about legal responsibilities. The job of a director or senior manager is to manage. The boardroom has the influence, power and resources to take initiatives and to set the pattern. So far as the first of our prerequisites is concerned — awareness — the cue will be taken from the top. We

know of a number of firms where the positive attitudes of the directors and senior managers are reflected in a remarkable degree of safety awareness at all levels throughout the firm. Conversely, if directors and senior managers are unable to find time to take a positive interest in safety and health, it is unrealistic to suppose that this will not adversely affect the attitudes and performance of junior managers, supervisors and employees on the shop-floor. If, as we believe, the greatest obstacles to better standards of safety and health at work are indifference and apathy, employers must first look to their own attitudes. Moreover, boardroom interest must be made effective. Good intentions at board level are useless if managers further down the chain and closer to what happens on the shop-floor remain preoccupied exclusively with production problems.

47. This takes us to the second and third of our prerequisites. The promotion of safety and health is not only a function of good management but it is, or ought to be, a normal management function—just as production or marketing is a normal function. The effective exercise of this function, as any other, depends upon the application of technique. Too many firms still appear to regard accidents as matters of chance, unpredictable and therefore not susceptible to "management". Too few appear to have made serious efforts to assess the total problem, to identify the underlying causes, or to quantify the costs. Too few make use of diagnostic and predictive techniques such as safety sampling or hazard analysis, or safety audits in which each aspect of workplace organisation and operation is subjected to a carefully planned and comprehensive safety survey; or systematic preventive procedures such as clearances for new equipment and processes, safe access permits and so on.

48. ... the employer who wants to prevent injuries in the future, to reduce loss and damage, and to increase efficiency, must look systematically at the total pattern of accidental happenings—whether or not they caused injury or damage—and must plan a comprehensive system of prevention rather than rely on the ad hoc patching-up of deficiencies which injury-accidents have brought to light. Shorn of the fashionable jargon used in much of the literature, this may seem no more than common sense. Yet it remains the case in industry that preventive action typically tends to take place in piecemeal fashion and only after the event.

Thus, management should firstly define clear policy statements within the firm; secondly, they should identify who is responsible within the management structure for the implementation of such policies. The following is from the same report:

53. ... It is generally accepted that the primary operational responsibility for ensuring safe working must rest with line management, and here there are two key levels that require particular attention. At the top, at board level, direct responsibility for the general oversight of safety and health matters within the firm should be included in the duties of one of the directors, in the same way that a director may be allocated overall responsibility for production or marketing or exports. In our investigations we formed the impression that undivided line management responsibility for safety and health matters more often than not stops at some point in the middle-management chain; further up the chain the responsibility tends to become diffused and uncertain. Safety and health should be treated like any other major management function, with a clear line of responsibility and command running up to an accountable individual at the very top. The other crucial level is the level of first-line supervision. It is the supervisor who is on the spot and in a position to know whether or not safety arrangements are working in practice. His influence can be decisive. Both here and abroad, wherever we have seen outstanding safety and health arrangements it has been

clear that a key role is played by well-trained supervisors who are held accountable for what happens within their sphere of control. We are not at all satisfied that this key role in safety is sufficiently recognised throughout industry generally, or that enough is done to equip supervisors for it.

THE ROLE OF THE SAFETY ADVISER OR OFFICER

The safety adviser, or officer, is employed by management, and thus shares the responsibilities of management in the promotion of accident prevention. He or she must be able to advise line management on all matters related to occupational safety and health. Normally, such a person has received specialist training in safety and is often placed in charge of safety throughout the firm. In short, the role of the adviser, or officer, is to supervise and promote accident prevention.

SAFETY COMMITTEES

For many years the trade unions have fought for the right to participate effectively in decisions on health and safety in the workplace. One of the main features of this participation has been the introduction of joint worker/management committees.

Safety committees are established to promote safety by co-operation between employer and workers; in some countries, the law makes their establishment compulsory, while in others they are voluntary. Management often uses the safety committee to explain the safety policy of the undertaking, for through the committee members it can reach all workers; conversely, the latter should turn to the committee to put their views and suggestions on safety matters to the management. The safety committee should help to give workers confidence in the safety policy of the management and help to make the management appreciate the safety experience of the workers. In brief, the safety committee should contribute to mutual understanding between management and labour and should encourage good team work.

Safety committees should consist of representatives of both the employer and the workers. The employer's representatives should always include the staff members who have direct responsibility for safety.

As many people as possible should serve on these committees, because membership is very likely to stimulate interest in safety. In some cases, worker representatives have been appointed for a limited period (e.g. one year), so as to allow all the workers to serve on the committee in turn. In most cases, the workers' members will be elected by them, but sometimes they will be nominated by the employer. When there are a number of departments, there may be a safety committee for each department and a central committee for the whole undertaking.

To increase the number of workers participating in committee meetings, one or two workers, who are not members of the committee but work in a department that is directly concerned with an item on the agenda, may be invited to attend a particular meeting to give their opinions or make proposals. The same people may be invited to a second meeting to hear what action has been taken on their proposals.

In some undertakings, safety committees have been working successfully for many years and have built up good relations and co-operation between employers and workers. In others, committees have been less successful; after a reasonably good start, the members have found themselves at a loss for subjects to discuss and have lost interest. Investigations into circumstances that have had a favourable influence on the work of safety committees, or that have been responsible for their failure, have shown the following considerations to be of particular importance:

1. The items on the agenda of every meeting must be carefully prepared, so that, if necessary, the chairman can give the members clear guidance on every point that may come up. The discussion must never be allowed to become confused. In some factories, the first item on the agenda of every meeting is a communication from the management which conveys information of general interest on, for instance, new developments and plans. The general atmosphere of committee meetings can sometimes be improved if members are given an opportunity to discuss subjects of particular interest to them, even if those subjects are not strictly related to safety. Everything possible should be done to keep the committee interested in its work.

2. Committees have often failed because they have overlooked the fact that promoting safety is a matter not only of goodwill but also of competence. This does not mean that every member of a safety committee has to be a safety expert, but it does mean that a sufficient number of members should know enough about safety to enable the committee to perform its task competently.

3. Worker representatives should always feel free to express their opinions and should not fear that if they make criticisms their superiors will make life difficult for them. An employer, or senior managerial official, will seldom object to constructive criticism; but sometimes a supervisor may take discussions on safety in his or her department as personal, and it may require the joint efforts of the management and the committee to smooth matters out and so allay the fear of reprisals in the worker's mind.

4. A safety committee must feel that it is backed by the employer. This can be ensured by having a member of the management as its chairman, providing suitable facilities, such as a comfortable, well furnished room for meetings, allowing members time off to attend

them, supplying secretarial assistance, permitting members to visit any place in the factory when there is good reason to do so in the interests of safety, and perhaps in other ways.

5. The committee should meet regularly — for instance, once a month. One item on the agenda of every meeting should be discussion of any accidents that may have occurred since the last meeting. The discussions should be directed to determining the cause of each accident and working out measures to prevent its recurrence.

6. Committee members, accompanied by the safety engineer or another competent person, should make periodical inspections of the undertaking. In this way, they will see conditions for themselves and have opportunities of discussing them with the safety engineer; it is important that the members should understand why the different safety precautions are taken and be able to judge their practical value. Inspections may also reveal how safety measures are carried out, and members will be particularly interested to see their recommendations put into effect.

7. The committee should be consulted on all proposals for new safety measures so that those finally adopted will have the greatest possible support of both the employer and the workers.

8. If a proposal made by the committee is rejected by the management, the committee should be told the reasons for its rejection.

9. All necessary information, such as statistics, should be given to the committee members, not only to keep them informed of the general situation and the accident trend, but also to provide them with a sound basis for discussing improvements.

Members of safety committees have the general duty of promoting the co-operation of all workers in the attempt to improve safety standards. They should try to make sure that safety instructions are followed and to counteract indifference and passive resistance. The dislike that many people have of taking orders may be avoided when the orders are explained by fellow-workers who are not in authority but who are respected or liked.

Committee members should try to replace the attitude of "I can take care of myself" with "I am a fool to take unnecessary risks". The greater the number of workers contributing to making the safety regulations, the more likely the committee is to be successful. The co-operation of the committee is particularly valuable when workers have to be told about safety regulations, instructions, and so on — a task more difficult than it seems. It is not sufficient to give workers a booklet of safety instructions; the contents must be explained to them, several times if necessary, to ensure that they are understood. On the other hand, safety committee members are in a position to inform the management of the practical value of the instructions given, and of improvements which are likely to make them more effective.

In some undertakings, the chairman of the safety committee is authorised to invite a worker who has had an accident to explain how it happened so that, if it was partly due to a mistake or fault on his or her part, the latter can be pointed out. Whether such an arrangement can be recommended or not depends entirely on the personality of the chairman. If such cases are handled tactfully, an advantage will be gained in that the criticisms of the worker's behaviour will come, not from the supervisor, who is responsible for safety and who is also his or her direct superior, but from a more neutral person who is concerned solely with accident prevention.

Lastly, members of safety committees should not forget that one of their main duties is to report hazardous conditions to the safety engineer immediately, and not to wait until the next meeting to do so.

It must not be forgotten that safety representatives and joint worker/ management safety committees are not the only means of promoting effective participation by workers in health and safety matters. There are many examples, particularly in small firms, of safety meetings which are attended by all workers. This ensures the participation of all concerned.

JOB SAFETY ANALYSIS

Just as productivity can benefit from work study (or job analysis), so can safety benefit from job safety analysis. What is more, the two are intimately bound together; people studying work cannot ignore safety, and those studying safety cannot ignore productivity considerations.

Job safety analysis, whether carried out as part of work study or not, can do much to eliminate the hazards of a job. The analysis isolates every single operation in a job, examines the hazards of each, and indicates what should be done about them. Work permits, drawings and tools, the qualifications required by the worker doing the job, and the instructions and training that the worker needs, are examined.

One way in which work study can simplify job safety analysis is by eliminating unnecessary operations and simplifying complicated ones. It is well known, for instance, that large numbers of factory accidents occur during the handling of materials. If work study can reduce the number of operations in which materials have to be handled, it will *ipso facto* remove potential causes of accidents.

SAFETY INSTRUCTIONS

Another safety measure used in factories is the issue of safety instructions for handling materials, operating machines and other kinds of work. Such instructions cannot replace protective devices, but may be useful in supplementing them or when their installation is not practicable. For instance, instructions should be issued on the way in which hoisting chains and wire ropes are to be used, stored and examined, and on the maintenance of machines and other equipment.

Preparing instructions is not difficult; the real problem is their enforcement. The best way to ensure that rules are obeyed is through effective participation. This can be done through the safety committee, or, if there is no safety committee, by some other form of consultation, perhaps through the trade union. It must not be thought that the issue of rules and instructions dispenses with the need for constant supervision. Indeed, supervision is the only means of ensuring that rules are understood and obeyed.

DISCIPLINE

The responsibility of the employer has been mentioned on several occasions. The employer is responsible for the workers' environment and the way in which work is carried out. However, it has also been mentioned that sometimes safety rules are not observed, and safety appliances are not used by the workers. If this occurs because the rules are not suited to the circumstances, there is no special reason to criticise the workers' behaviour. It is another matter, however, if a worker thinks that guards are unnecessary and therefore does not use them. Here, normal factory discipline has to be enforced; if such behaviour endangers the worker or other workers, disciplinary measures will have to be taken. It should be pointed out that discipline in this context does not mean punishment as such, but something far more positive and constructive. The worker must be trained to follow the correct safety procedures.

INTRODUCTION OF NEW WORKERS

Special care should be taken over introducing new workers to the undertaking. They should be acquainted with the new environment and told what is expected of them.

One of the reasons why statistics show a relatively high number of accidents among new workers is doubtless because they are not educated and trained properly. All too often, new workers, when arriving at the factory for the first time, are sent immediately to the department where they are to work and have to wait there until someone has time to show them their place and their work. They are told a few things, and then they are left to their own devices.

The impression new workers receive from the factory in general, and from their own environment in particular, is very important. Whether they will be interested in the factory and their fellow-workers, or only in their wages, and whether they will feel that they are full members of a new society, or only outsiders, will depend to a great extent on the impressions they gain during the first few days.

New workers should be shown generally how the factory is run and should be given an opportunity to ask questions. They should be given information on subjects of special interest to them, such as wages, working hours, the canteen, the first-aid service and welfare facilities. The department in which they are to work should be described in greater detail, and there should be full explanations of the working methods that they have to follow. Safety instructions should be explained clearly and fully, and great care should be taken to ensure that the workers really understand them.

The importance of obtaining immediate first-aid treatment after an accident should also be stressed, so that workers are encouraged to report immediately to the first-aid service after any injury, however minor. Something should be said of the importance of good order and good housekeeping. It is most important that newcomers be introduced to their supervisors and to fellow-workers.

The methods used to introduce new workers will differ from one place to another. The introduction may take place during a talk in the personnel department, or in the safety engineer's office, or even in the factory department concerned, and may be conducted by a member of the personnel department, the supervisor or an experienced worker. In some cases, the introduction includes an explanation of the technology of the undertaking, and a plan or model is shown to make things clear. In other cases, illustrations are provided by means of a film-strip or a film.

Although it may be necessary, for formal reasons, to provide new workers with a copy of the factory regulations and instructions, it cannot be expected that they will fully understand them, or even that they will read them carefully. During the introductory talk, those regulations and instructions which are of immediate concern should be explained. After the explanation, the correct way of working should be shown and the workers should be allowed to try it themselves, to make sure that they have understood it. Subsequently, regular checks should be made to ensure that they are working in the prescribed way, and that no departures from safe working habits have occurred.

The main purpose of such introductions is to make personal contact with the newcomers and to show a personal interest in them. If the person in charge of an introduction considers the procedure only as a boring routine, it will do more harm than good. Workers should be made to feel that, if they encounter an unexpected difficulty in the course of their work, there is somebody to whom they may go for information and advice; this will help them to feel at home in the factory, and also to feel something like job satisfaction, which is necessary for the peace of mind that reduces the risk of accidents.

Some firms arrange for another meeting with newcomers, about two or three months after their arrival, to have a talk about their experiences, needs and desires. On this occasion, difficulties which have arisen can be considered, and unsatisfactory conditions can be corrected. In such ways, labour turnover, which is also important from the safety point of view, may be reduced.

QUESTIONS

1. To what extent does safety in a factory depend on the management?

2. What reasons are there for believing that safety committees are useful?

3. Why are some safety committees failures?

4. How should new workers be introduced into the factory?

EARLY INDUSTRIAL SAFETY ACTIVITIES OF GOVERNMENTS, PUBLIC AUTHORITIES AND PRIVATE ASSOCIATIONS

12

In this lesson, we shall examine some of the early industrial safety activities of governments, public authorities and private associations.

The Prevention of Industrial Accidents Recommendation, 1929 (No. 31), contains a comprehensive statement of the principles that should govern the safety activities of governments, public authorities, industrial associations, insurance institutions, other bodies, employers and workers.

The functions assigned to governments and other public authorities are substantially as follows:

(a) the collection and use of information on the causes and circumstances of accidents;

(b) the study, by means of statistics of accidents in each industry as a whole, of the special dangers that exist in the several industries, the "laws" determining the incidence of accidents and the effects of measures taken to avoid them;

(c) the carrying out of methodical investigations, where appropriate with the assistance of institutions or committees set up by individual branches of industry;

(d) investigation of physical, physiological and psychological factors in accidents;

(e) encouraging scientific research into the best methods of vocational guidance and selection, and their practical application;

(f) establishing central departments to collect and collate statistics relating to industrial accidents;

(g) developing and encouraging co-operation between all parties interested in the prevention of industrial accidents, and particularly between employers and workers;

(h) arranging for periodical conferences between the state inspection service, or other competent bodies, and representative organisations of employers and workers in every industry, or branch of industry, to review the accident situation and discuss proposals for improving it;

(i) encouraging the adoption of safety measures such as the establishment of works safety organisations, co-operation between management and workers in individual works, co-operation between employers' and workers' organisations in an industry, co-operation between these organisations and the State and with other appropriate bodies;

(j) awakening and maintaining the interest of the workers in the prevention of accidents and ensuring their co-operation by means of lectures, publications, films, visits to industrial establishments and other appropriate means;

(k) establishing, or promoting the establishment of, permanent safety exhibitions;

(l) ensuring, by indirect means, that employers do all in their power to improve the education of their workers in regard to the prevention of accidents, and that workers' organisations should co-operate in this work;

(m) arranging for handbooks on accident causation and prevention in particular industries, or branches of industry, or particular processes, to be prepared by the state inspection service or other competent authorities;

(n) arranging for the inclusion, in the curricula of primary schools, of lessons designed to inculcate habits of carefulness and, in the curricula of continuation and vocational schools, of lessons in accident prevention and first aid;

(o) prescribing by law the measures required to ensure an adequate standard of safety;

(p) examining plans for the construction of, or substantial alteration of, industrial establishments;

(q) consulting representative organisations of employers and workers before issuing administrative orders or regulations for the prevention of accidents;

(r) providing for the collaboration of the workers in securing the observance of safety regulations;

(s) endeavouring to secure that accident insurance institutions or companies take into account, in assessing the premium for an undertaking, the measures taken in it for the protection of the workers;

(t) inducing accident insurance institutions and companies to co-operate in the work of accident prevention.

This is a fairly comprehensive programme of activity for the State, but it is not all the State should do. It will be noticed that little is said about the extremely important subject of the organisation and functions of the state inspection service. These are dealt with in the Labour Inspection

Recommendation, 1923 (No. 20), and in the Labour Inspection Convention, 1947 (No. 81), and Recommendation, 1947 (No. 81). We shall return to these subjects later in this lesson.

The functions attributed to governments and public authorities in these international instruments will be seen to fall into various classes: safety laws and regulations, enforcement investigation, education, research and testing, promotional activities, and so on. These different branches of activity are described below.

SAFETY LAWS AND REGULATIONS

The Prevention of Industrial Accidents Recommendation, 1929 (No. 31), states that any effective system of accident prevention should rest on a basis of statutory requirements. This statement is amply justified by experience.

There are many kinds of safety laws and regulations. In countries with parliamentary systems of government there is a clear distinction between a law and a regulation. A law is passed by parliament; a regulation is issued by a minister, although it may have to be submitted to parliament before it can become effective. Passing a law through parliament is usually a very laborious process, and consequently there is a tendency in many countries to draft safety laws in such a way that they do not frequently have to be resubmitted to the legislature for amendment in the light of subsequent developments. This is done by confining such laws to general principles and including a list of subjects on which the competent minister has the power to issue more detailed regulations. For instance, the United Kingdom Factories Act of 1937 contains a provision which reads as follows:

60 (1). Where the Secretary of State is satisfied that any manufacture, machinery, plant, process or description of manual labour used in factories is of such a nature as to cause risk of bodily injury to persons employed in connection therewith, or any class of those persons, he may, subject to the provisions of this Act, make such special regulations as appear to him to be reasonably practicable to meet the necessity of the case.

Many regulations have been issued under the United Kingdom Factories Act. Among the subjects with which they deal are lighting, chemical works, the manufacture of aerated waters, the blasting of castings, eye protection, the manufacture and decoration of pottery, cotton-spinning, woodworking machinery, the lifting of heavy weights, spray-painting, electricity, lifting machines, operations at unfenced machinery, milling machines, grinding of metals, plant railways, acetylene, the manufacture of celluloid, the grinding of magnesium, luminising and the training of young persons to operate dangerous machines. This short list is enough to show how complicated safety legislation is becoming.

Indeed, the factory safety regulations of quite a number of countries would make up substantial volumes (say, 500 pages of a book like this one).

Although the amount of safety legislation is increasing rapidly, laws and regulations by themselves can never be enough to achieve the highest attainable standard of safety. They can only embody provisions that are enforceable without too much difficulty, and that can be applied by all those responsible for applying them. This means that they have to be as simple as possible, and not too difficult for those undertakings with the very lowest standards of safety in the country to apply. Consequently, as a rule, laws and regulations lay down a bare minimum standard of safety. Moreover, they do not usually provide guidance as to how this standard is to be achieved.

Laws and regulations often have to combine technical with legal phraseology, and, as a result, they are sometimes not easy for the layman to understand. A useful practice followed in some countries is to issue booklets explaining the provisions of safety laws in simple language.

Here are one or two examples of legislative provisions, illustrating some of their limitations.

In the Netherlands, the following regulation is in force: *Wooden ladders should not be painted, but coated with clear varnish.* This rule indicates exactly what has to be done, but does not explain that it should be done to prevent defects in the wood from being concealed by paint.

Mexican legislation contains the following: *If it is necessary to nail anything, use should always be made of a hammer and not of any object, because if the tool is not suitable it is likely that the blow will glance off the head of the nail and fall on the worker's fingers.* Here a safety rule is formulated, and the need for such a rule is explained as well.

A New York State regulation reads as follows: *Each extractor hereafter installed and existing installations, where practicable, shall be equipped with an interlocking or other approved device that will prevent the cover being opened while the basket is in motion and also prevent the operation of the basket while the cover is open.* In this case the purpose of the guard is indicated, but no information is given on how to install the guard or why it is necessary.

Safety regulations often use general terms which require an explanation to make their intention clear. For instance, they may lay down that "measures should be taken" without indicating what measures; that "suitable" guards should be used without mentioning what, in that particular case, "suitable" means; that precautions have to be taken "in good time" without fixing a period; that measures are not necessary when machine parts are "safe by position" without prescribing in what circumstances the position can be considered to be safe. In practice, the changes which are continually taking place in many sections of industry make it impossible to avoid such vague general expressions. To mention all the details of guards in a regulation would result in a very unwieldy document, and, even if it were possible to compile it, the regulation might well prove a serious impediment to new ideas on safety and new developments in safety engineering.

In some countries, to meet the difficulties created by wording regulations in general terms, the legislation names an authority (e.g. an official of the labour inspection service) which is empowered to prescribe in a particular case what the employer has to do to comply with a certain regulation. In other countries, there is an authority responsible for approving guards. In yet others, the decision as to whether a safety precaution is in accordance with the law is a matter for the courts, which usually decide after hearing the evidence of safety experts. In any case, some authority must be in a position to give a valid interpretation of safety regulations.

In industrially advanced countries, safety regulations have been gradually built up over a long period. These countries have acquired much experience of the practical value of safety regulations and of the manner of enforcing them. In some industrially developing countries, where the necessity for safety legislation dates from only a few years back, the situation is very different; in those countries, the regulations are sometimes based on those of the ILO *Model code of safety regulations for industrial establishments*, which in turn are based on the experience of the industrialised countries. However, the difference between the two groups of countries, is that, in the industrially developing ones, there are often very few persons (if any) who have experience of safety regulations, or who are familiar with their history, and this difficulty has proved to be almost insurmountable in the enforcement of the regulations.

A solution to this problem has been sought by sending a number of persons with a suitable technical background from these countries to study safety questions in countries with more experience, or by sending experienced advisers to the industrially developing countries to adapt the regulations to local conditions and to pass on experience acquired elsewhere. To a certain extent, too, the lack of knowledge and experience can be remedied by means of manuals.

Besides safety laws, laws dealing with accident insurance also contribute to safety. In many countries, employers have been made liable by law to pay compensation in cases of accidents; in some, the law obliges employers to reimburse insurance costs if an accident is due to inadequate protection, while, in others, the insurance institute increases or decreases premiums according to the safety precautions taken in an undertaking. In all these cases, the employer has a direct financial interest in ensuring safe working conditions, and, to this extent, the law promotes safety.

Accident insurance legislation has much more influence on safety when it authorises insurance institutions to issue safety regulations and to supervise their enforcement. In this way, the administration of accident compensation and prevention are combined in one institution. This is useful, since the accident prevention department of the insurance institution is immediately informed of accidents. One advantage of having safety work done by social insurance institutes is the ease with which funds can be provided for research, educational and other work. It is a sound policy to invest money in this way, for it will yield a high return in the form of a reduction in the number of compensatable accidents.

ENFORCEMENT OF LAWS AND REGULATIONS

Not very long after the first safety legislation was passed, governments realised that it would be ineffective unless some means could be found of enforcing it. After some experimenting with different enforcement agencies, the great majority of countries eventually established state labour inspectorates, factory inspectorates or similar enforcement agencies. In some countries, however, this function is wholly or partly entrusted to other bodies, such as trade unions or associations of boiler owners.

Reference has already been made to the general principles for the organisation of labour inspectorates set out in a Recommendation adopted by the International Labour Conference in 1923.[1] On the specific subject of safety, the Recommendation includes the following provisions:

(a) ... one of the essential duties of inspectors should be to investigate accidents ... with a view to ascertaining by what measures they can be prevented;

(b) that inspectors should inform and advise employers respecting the best standards of health and safety;

(c) that inspectors should encourage the collaboration of employers, managing staff and workers for the promotion of personal caution, safety methods, and the perfecting of safety equipment;

(d) that inspectors should endeavour to promote the improvement and perfecting of measures of health and safety. . . .

The provisions concerning the organisation of inspectorates include a recommendation that, in view of the difficult scientific and technical questions which arise under the conditions of modern industry, experts having competent medical, engineering, electrical or other scientific training and experience should be employed by the State.

The Recommendation further states that inspectors should possess a high standard of technical training and experience, be persons of good general education, and by their character and abilities be capable of acquiring the confidence of all parties. One important point made is that inspectors should be given such a status and standard of remuneration as to secure their freedom from any improper external influences.

With regard to standards of inspection, it is suggested that, as far as possible, every establishment should undergo a general inspection at least once a year, and that unsatisfactory or particularly dangerous or unhealthy establishments should be inspected much more frequently. The Recommendation stresses the importance of co-operation between the inspectorate, employers and workers.

Labour inspection services [2] usually have to supervise the enforcement of all labour protection laws, such as those dealing with safety, health, working hours, rest periods, the protection of female and child labour and minimum wages. In addition, they have to advise employers and workers on social questions and keep the government informed of social conditions in the country.

Normally, the inspection service has a headquarters staffed with experts in different fields and with administrative personnel, and the country is divided into districts, each one under a district superintendent, assisted by district inspectors and administrative personnel.

The district inspectors make the factory inspections, including safety inspections, under the general guidance of the district superintendent. They should be sufficiently familiar with the safety laws and regulations to be in a position to explain them and to advise on the best ways of complying with them. If they themselves require information on technical questions they may ask the district superintendent, who, if necessary, may refer the matter to the central authority. With its specialists in medicine, chemistry, electricity, engineering, etc., the central authority will be in a position to assist the labour inspectors, and to provide them with any information they require.

Even inspectors with good technical knowledge are not in a position to deal with all the safety problems which arise in industry. Separate inspection services have been organised in most countries for boilers and mines, and in some countries also for lifts, ports, agriculture, electricity, and other matters.

The specialisation necessary to make accident prevention work effective, in the face of the ever-increasing complexities of technology, threatens to overload labour inspectorates. The desire to avoid this, and at the same time to take advantage of private facilities, has sometimes led governments to make private associations (e.g. associations of manufacturers of boilers, of acetylene installations and of lifts) responsible for safety inspection in their particular field.

In some countries, the low salaries offered to inspectors seriously hinder efficient inspection, since it is impossible to recruit enough qualified personnel. Inspectors may be appointed who lack technical knowledge and who do not themselves understand the safety regulations they are supposed to enforce. Such inspectors can scarcely be expected to be good advisers.

It is also undesirable that inspectors should be dependent on travelling facilities provided by employers or that they should receive their fees from employers; such a situation makes it difficult for them to act with the necessary independence.

It is most important that inspectors should be technically competent and command the respect of the people with whom they have to deal. An inspector with inadequate technical knowledge, or not well acquainted with factory conditions, who recommends impracticable or unsuitable safety measures, is a nuisance to employers and workers alike, and does little to prevent accidents. This sort of situation is liable to arise in countries in which industry is in an early stage of development.

In cases of imminent danger, labour inspectors are sometimes authorised to order that work should be stopped. As this is a very radical measure, the power to stop work should be used with the utmost caution. For this reason, in some countries, stoppage of work can only be ordered by a court on the application of a labour inspector. In other countries, an order may be given by the labour inspector, but it has to be confirmed by a court within a prescribed period.

Example. An imminent danger exists when a welding torch is used for repairing a petrol-tank truck, without measures being taken to prevent an explosion of the mixture of petrol vapour and air inside the tank. Work should be interrupted immediately to eliminate the petrol vapour, or to fill the tank with an inert gas, such as nitrogen or carbon dioxide.

Example. The hoisting of materials above a place where men are working creates an imminent danger of injury from falling objects. As long as the men have to be at that place, no hoisting should be performed.

EDUCATIONAL AND ADVISORY WORK BY STATE SERVICES

Safety programmes, courses and demonstrations have also been developed by state services, in particular for the smaller undertakings in some branches of industry. For instance, the Bureau of Labor Standards of the United States Department of Labor has prepared flow charts showing the sequence of plant operations with the principal hazards. A six-month safety programme has been devised and assistance is given in carrying it out. At the end of the period, the results are assessed by calculating accident rates and by comparing the figures with the situation before the programme began. There is a follow-up programme to continue the safety activities. [3]

Figure 50 shows an inspection sheet for a bale-breaker in the cotton industry, used in the safety programme. The points to which special attention should be paid in the interest of safety are as follows:

1. *Operator.* Stands to one side when cutting bale-ties, not in front of tie being cut. Wears gloves and safety glasses when handling bale-ties. Uses tie cutter to cut bale-ties.
2. Motor frames grounded.
3. Sufficient light without glare.
4. If overhead line shaft drive, mechanical belt-shifters provided.
5. Belts and pulleys fully enclosed.

6. *Operator*. Stops machine before oiling or cleaning.
7. Inspection panel. Operator does not open while machine is in operation.
8. Machine firmly anchored.
9. Guard covering exposed gears.
10. Floors free of oil, ties, tie buckles or obstructions, etc.
11. Spiked apron or beater covered. Operator does not open cover while machine is in operation.
12. *Operator*. Rake provided and used for cleaning out motes.
13. Protruding shaft ends or revolving shafting guarded.

Figure 50.   An inspection sheet for a bale-breaker in the cotton industry

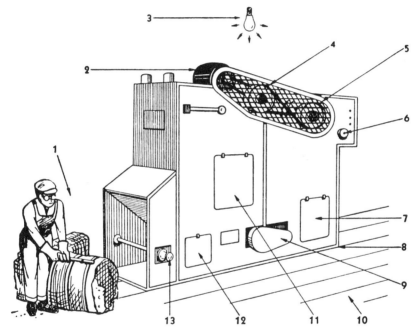

A very important contribution to accident prevention is made by the reports, periodicals, leaflets, posters, etc., published by labour inspection services and insurance institutes. These publications contain excellent material for the student of industrial safety.

RESEARCH AND TESTING

The vast development of technology has led labour inspectorates and other state services to set up laboratories for the analysis of dangerous chemical substances, for testing materials and equipment (see Lesson 2), and for research on dangerous working methods. Inspection services have acquired much experience in this field, and they have done valuable research work.

Example. When overhead recessing machines were introduced in the woodworking industry, they were either completely unguarded or provided with clumsy guards at the point of operation. Serious accidents happened to workers who occasionally came into accidental contact with the tool, because of an unsuitable working method, or as a result of workpieces being kicked back or thrown out by the machine. To remedy this situation, the Swiss National Accident Insurance Institute (SUVA) undertook research, studied the different types of work done on these machines, and developed a guard, with the assistance of operators and heads of undertakings. Subsequently, some manufacturers fitted their new machines with these guards, thus testifying to the value of the work performed by the Institute.[4]

The testing of materials from the safety standpoint is often done by inspection services in co-operation with specialised laboratories, for the inspection service itself may not be equipped for the purpose. Tests are important as a means of discovering the cause of an accident in which a broken chain, wire rope, rod or other piece plays a part. They enable conclusions to be drawn as to whether the cause of the accident was the use of the wrong material, overloading, or treating a material wrongly (e.g. incorrect heat treatment of steel). In some cases, the co-operation of the boiler inspection service may be requested for the investigations; in others, the inspectorate may turn to a mining research laboratory, the laboratory of a technical university or a specialised private laboratory.

Machine guarding is a field that has benefited greatly from the work of testing laboratories. Testing has prevented the introduction of unsuitable guards and has yielded valuble information on the structural details of these and other appliances. Good results have also been achieved through the testing of grinding-wheels, chains, wire ropes, structural parts of scaffolding, ladders, safety shoes, etc.

Some private laboratories, which test materials and examine constructions and apparatus, make important contributions to industrial safety, particularly when they concentrate on such work. For example, in the laboratories of the Association of Belgian Manufacturers *(Association des Industriels de Belgique)*, investigations are made on boilers, hoisting appliances, centrifugal driers, welding installations, electrical installations, chains, wire ropes, safety belts, ladders, etc. They have the necessary equipment for testing materials in different ways and for chemical and metallographic analysis.

PRACTICAL WORK BY STATE SERVICES

A very interesting practical contribution to safety in the woodworking industry has been made by the Swiss National Accident Insurance Institute (SUVA). After designing guards first for circular saws and moulding machines, and later for planers and overhead recessing machines,[5] the Institute not only made them available to industry by organising their manufacture and distribution, but also provided highly qualified instructors to demonstrate how the guards should be used under the most varied conditions. In this way, the workers learned not only how to perform dangerous operations safely, but also how to improve the quality and quantity of production. This gesture has proved to be extremely useful in limiting the number, and in particular the severity, of woodworking-machine accidents. The example has so far been followed only by the Netherlands.

Another service to industry, provided only in the Netherlands and in Switzerland, is the provision of assistance in the mounting of certain guards. In both countries, guards for power presses have been developed which require very accurate mounting to be efficient. To ensure that they are not rendered ineffectual by faulty mounting, officials of the insurance institute or labour inspection service attend to give advice when they are fitted.

CO-OPERATION BETWEEN INSPECTORATES AND EMPLOYERS AND WORKERS

It has already been seen that, although labour inspectors have to enforce laws and regulations, they should be more than merely a kind of technical police force; they should also be able to give advice on safety and health matters. As accident prevention is of interest not only to the workers but also to the undertaking, and indeed to the country as a whole, many safety problems can best be solved by co-operation between inspectors, employers and workers. Much more can be achieved by co-operation than by the bare enforcement of the law, for, as we have said, the law embodies only minimum precautions.

Accordingly, labour inspectors can contribute to safe working conditions by giving advice based on their technical knowledge and experience, as well as by simply knowing the regulations. Inspections of factories, and investigations of accidents, afford useful opportunities for discussions of safety matters with the management, supervisors and workers, from which all can benefit. However, as there are far fewer inspectors than there are establishments, it is obvious that a particular factory will not be visited by an inspector very frequently. Safety in factories cannot be ensured by labour inspectors alone.

STATE-OWNED INDUSTRIES AND GOVERNMENT CONTRACTS

Another means by which governments can promote safety lies in the existence of state-owned or state-controlled industries. In these industries, the standard of safety could be brought up to a level that would make them models for private industry.

Government contracts with private firms for the construction of roads and buildings, the delivery of machines, etc., provide governments with opportunities for stipulating the safety measures that have to be taken in carrying out the contract.

SAFETY MUSEUMS AND EXHIBITIONS

Safety museums may be said to have originated in the international safety congresses held at the end of the nineteenth century. A new way of propagating knowledge of safeguards was proposed at the Milan Congress (1894), although it had been mentioned still earlier at a hygiene congress held in Vienna (1887). The proposal was to establish "social museums" where the public could see models and obtain information on social insurance and, in particular, on accident prevention.

The original purpose of safety museums was to show the best and most modern means of protecting workers against the risks of their work (for instance, specimens of effective guards for different machines, of personal safety equipment and of other means of preventing accidents and disease). In practice, this aim was never completely realised. The earliest safety museums were partial failures, for there was not enough money to keep them fully equipped and up to date.

There were other difficulties. As was mentioned in Lesson 9, only a relatively small number of people visited the museums, and, among them, only a few could be considered as being directly interested in safety problems. Moreover, demonstrations of guards in a museum did not convince some people that they would also be suitable under factory conditions.

Some museums have continued along conventional lines, but they are not very important today. Others have been reorganised and developed into institutions which now take an active part in the safety movement. One important change was the abandonment of the practice of inviting employers and workers to visit safety museums and the organisation of exhibitions in factories instead. Itinerant exhibitions of this kind have been organised and placed at the disposal of industry. It has thus become possible to provide an undertaking with what it needs at a particular moment for its safety programme. Similarly, films, film-strips and slides are no longer used solely for demonstrations in museums but are also available to factories. Safety museums sometimes open lending libraries and information centres as well.

Lastly, members of the staff of some safety museums are now sometimes loaned to undertakings, and especially to smaller ones. They may go to a factory and stay there for a month, or longer, to organise safety activities, to make inspections and to advise generally on accident prevention. This has proved to be an excellent method when a safety engineer on the staff of the undertaking would be too costly or could not be fully employed. Guards may also be borrowed from the museums so that their practical value can be demonstrated to employers and workers.

SAFETY ASSOCIATIONS

So far in this lesson we have been mainly concerned with the safety activities of the State — the legislature, ministries, technical institutes, and so on. In many countries, these agencies do not handle all safety work, and much may be done by private bodies of various kinds — safety associations, standards institutions, employers' associations, trade unions, universities, private laboratories, and others. However, it should not be forgotten that the functions exercised by voluntary bodies in some countries are exercised by state agencies in others.

There are different kinds of safety associations. Some cater for all or most industries, and for the whole country; examples are the National Safety Council of Australia, the Royal Society for the Prevention of Accidents in the United Kingdom, and the National Safety Council in the United States. Others, such as the Normandy Industrial Safety Association in France, the Alberta Safety Council in Canada, and the Antwerp Provincial Safety Institute in Belgium, have a similar coverage of industry but are regional bodies. Yet others deal exclusively with a particular industry or a particular subject; examples are the Construction Safety Association of Ontario (Canada) and the Trade Association for Accident Prevention in Building and Civil Engineering in France.

Then there are industrial associations that are not wholly safety associations but have safety departments or engage in a wide range of safety activities. Examples are the Association of British Chemical Manufacturers, the Associated General Contractors of America, and the Japan Coal Association.

Some large insurance companies, such as the Employers' Mutual Liability Insurance Company of Wisconsin (United States), also have safety departments.

Other examples of various kinds of association that engage, in one way or another, in safety activities are the Ahmedabad Textile Industry's Research Association in India, the American Society of Safety Engineers, the Algerian Association of Owners of Steam and Electric Plant, the British Colour Council, the Italian Centre for Safety Officers, the Safety Engineering Society, Melbourne (Australia), the German Federation of Electrical Engineers, and the Association of German Inspection Engineers.

The activities of these associations are as varied as their forms. It will scarcely be possible here to describe all the safety activities of all the types of safety association; but the types of work that they carry on include the following:

1. Organisation of national safety congresses.
2. Organisation of safety exhibitions.
3. Training of safety officers and other persons responsible for safety.
4. Operation of libraries.
5. Operation of information services.
6. Organisation of special safety campaigns.
7. Practical assistance to small firms.
8. Compilation of safety codes and rules.
9. Publication of technical literature.
10. Publication of periodicals.
11. Publication of leaflets, posters and other propaganda material.
12. Organisation of publicity through the press, radio, television, etc.
13. Compilation of accident statistics.
14. Plant inspections.
15. Testing of materials, substances, atmospheres, etc.
16. Technological (physical and chemical) research.
17. The granting of awards to holders of safety records, winners of safety competitions, performers of meritorious services, etc.
18. Collaboration with state agencies such as the labour inspectorate.
19. Collaboration with institutions working in the safety field or related fields, such as standards, health, personnel management, scientific research, educational and welfare institutions.

STANDARDISATION

In assessing the contribution to safety made by private associations, a special place should be given to standards associations. Although these associations do not work exclusively in the interests of safety, they contribute substantially to it. Some results of standardisation have already been mentioned in connection with accident rates in Lesson 3. The formulation of general rules for the calculation of these rates has made it possible to compare the accident situation in different factories in the same country.

Standards associations were at first mainly concerned with reducing the unnecessary and inconvenient variety of common technical objects such as rivets, bolts and nuts; but their fields of activity soon expanded, and standardisation was applied to material and construction specifications, testing materials, design calculations, etc.

Nowadays, many kinds of standards are published that have a bearing on safety. Some are actually called safety standards, codes of practice, etc., that deal to some extent with safety or have implications in the safety field.

The following are examples of various subjects of safety interest dealt with in national standards:

(a) *industrial equipment:* ladders, abrasive wheels, pressure piping, boilers, elevators;

(b) *personal protective equipment:* glasses, respirators, gloves, hats, boots, aprons;

(c) *colours, signs, signals, symbols:* identification of piping systems, identification of gas cylinders, safety colours, signalling devices for printing presses, hand signals for hoisting appliances;

(d) *safe practices:* safety procedures for quarries, industrial uses of X-rays, precautions against fire, installation and maintenance of flame-proof and intrinsically safe electrical equipment; and

(e) *accident records:* recording and measuring work injury experience, compiling industrial injury rates.

Standards are valuable largely because they are the result of the co-ordinated efforts of all the parties concerned (manufacturers, users, scientists, etc.). Since safety standards enjoy universal approval, they have on several occasions been incorporated in official regulations. For instance, there are regulations which require the material used for crane hooks and slings to conform to the national standard, prohibit the use of respirators other than those conforming to the national standard, or allow only the use of standard electrical appliances.

One great advantage of standardisation is that it protects manufacturers of approved standard products against complaints by the user or the inspection service; moreover, users know precisely what they are buying and can thus avoid having to make expensive changes or adjustments in equipment.

As international contacts steadily multiply, and the advantages of international standards in international trade become more apparent, greater efforts are being made in the direction of international standardisation through the International Organisation for Standardisation (ISO), and the International Electrotechnical Commission (IEC), under which the International Commission for Conformity Certification of Electrical Equipment (CEE) operates. International standards drawn up by the ISO include codes of safety on roll-over protection for tractors and earth-moving machinery, identification of hazard points in ground handling of aircraft, and guard rails for cargo ships. The IEC, which is responsible for international standardisation in the electrical and electronic engineering fields, has drawn up a large number of safety standards, while the CEE, to which about 15 European countries belong, has framed a considerable number of recommendations concerning the safety of electrical apparatus and materials.

Standardisation can become an important safeguard to ensure that only safe appliances and machines are sent to technically developing countries. In some countries, two types of machines are manufactured: one well guarded and destined for countries with advanced safety requirements, and one without guards for the others. This practice should be abandoned, but it will continue until there are world-wide safety standards.

QUESTIONS

1. Mention some of the things that governments have done to promote occupational safety.

2. Mention some of the ways in which laws and regulations have been *(a)* effective; and *(b)* ineffective in promoting safety.

3. What measures have governments taken to get safety laws and regulations obeyed?

4. What requirements should a state labour inspection service satisfy?

5. Why is co-operation between the labour inspection service, employers and workers essential?

6. Name some uses of safety exhibitions.

7. What kinds of work can voluntary safety associations do?

8. In what ways can standardisation contribute to safety?

Notes

[1] The main general provisions of this Recommendation were subsequently embodied in a Convention (the Labour Inspection Convention, 1947 (No. 81)).

[2] See also ILO: *Labour inspection: Purposes and practice* (Geneva, 1973) and the Labour Inspection Convention, 1947 (No. 81).

[3] See "Special industry safety programmes", in *Occupational Safety and Health* (Geneva, ILO), Jan.-Mar. 1954, p. 28.

[4] See "The safe operation of overhead recessing machines for woodworking", in *Occupational Safety and Health* (Geneva, ILO), Apr.-June 1954, pp. 63-68.

[5] ibid.

INTERNATIONAL SAFETY ACTIVITIES 13

International action for accident prevention did not develop on any considerable scale until nearly a century after national action had begun. It has followed the lines of national development, but in many respects it is still very incomplete.

BEGINNINGS

A modest beginning of international co-operation in accident prevention may perhaps be seen in one of the congresses to which we have referred several times, the International Industrial Accident Congress, held in Paris on the occasion of the Universal Exhibition in 1889. The Congress was organised by a committee of 30 persons with some concern for, or interest in, accident prevention and compensation, designated by a French Ministerial Order of 26 December 1888. The Congress, which was concerned both with prevention and with compensation, was divided into three sections, the first dealing with technical questions, the second with statistics and administration, and the third with economic questions and legislation.

The delegates recommended the establishment of a permanent international body to compile the experience gained in the different countries, and suggested the best procedure to be followed in the future. As a result, a permanent international committee was set up in 1890; one of its specific tasks was to try to find a basis for the compilation of international accident statistics. By 1891, it had a secretariat and 600 corresponding associates in different countries, and it aimed to become an International Industrial Accident Office.

A second International Industrial Accident Congress was held in Berne in September 1891; several papers on accident prevention were read. Other conferences of the same kind were held in Milan in 1894, Brussels in 1897 and Paris in 1900; at the latter, 18 countries were represented.

Another step forward in the international organisation of accident prevention may be seen in the foundation of the International Association for the Legal Protection of Workers in 1898. Under the auspices of this Association, an International Labour Office was established at Basle, which began the publication of an international bulletin of labour legislation in 1902; numerous safety regulations were published in this periodical. The Association itself also took an interest in industrial safety and accident statistics.

By the time of the Paris Peace Conference of 1919, industrial safety had assumed such importance that the authors of the Treaty of Versailles specifically mentioned in the preamble to its labour clauses that the protection of the worker against injury arising out of his employment was one of the measures urgently needed to improve conditions of labour.

THE INTERNATIONAL LABOUR ORGANISATION

The First World War was the end of an epoch, especially in Western Europe, which had enjoyed over 40 years of uninterrupted peace. As the war dragged on, the feeling gained ground that such horror and cruelty should never occur again, and that the restoration of peace should usher in a new era in which people should live in decent conditions and enjoy a measure of what was subsequently to be known as social justice. The working classes of many countries looked forward to better living conditions after the war, and governments promised that their wishes would be granted. At an international workers' conference, held in Leeds in 1916, various demands were formulated, including the following:

The different countries should bind themselves to develop their legislation on occupational hygiene and safety. They should try to unify their legislation for every branch of industry. In particular they should organise permanent co-operation for common action against industrial poisons, defective or dangerous manufacturing processes and occupational diseases.

At an international trade union conference, held in Berne in 1918, proposals for an industrial labour charter incorporating these demands were adopted, together with a demand for the reduction of working hours in dangerous industries. Other workers' congresses made similar demands.

The Treaty of Versailles put an end to the First World War as far as Western Europe was concerned. Part XIII of that Treaty established the International Labour Organisation. One reason for its foundation, given in the Preamble to Part XIII, was that it was urgently necessary to improve conditions of labour, and, in particular, the "protection of the worker against sickness, disease and injury arising out of his employment". A further argument put forward was that "the failure of any nation to adopt humane conditions of labour is an obstacle in the way of other nations which desire to improve conditions in their own countries".

The Treaty also considered it to be of special and urgent importance that "each State should make provision for a system of inspection in which women should take part, in order to ensure the enforcement of the laws and regulations for the protection of the employed".

When the ILO was founded in 1919, it had 42 member States. By 31 July 1972, the number had risen to 123, and by January 1983 to 150.

In 1919, the industrialised countries already had safety laws and regulations, with labour inspectorates to watch over their enforcement. In factories, employers were obliged by these laws and regulations to take certain precautions against accidents; in practice, however, safety standards left much to be desired. There may have been several reasons for this: safety regulations were not nearly so highly developed as they are now, nor labour inspectorates so well organised; many safety devices were impracticable because the technology of the processes in which they were to be used had not been thoroughly studied; and, more generally, public concern for safety was slight.

Structure and functions of the ILO

The permanent machinery of the International Labour Organisation essentially consists of the International Labour Office, its Governing Body and the International Labour Conference. For special purposes, such as the discussion of international standards, special conferences of workers', employers' and government representatives, known as tripartite technical conferences, may be held, and ad hoc committees of experts may be appointed by the Governing Body to examine technical problems.

The main functions of the International Labour Office in the field of occupational safety are:

(a) the preparation and revision of international standards (Conventions, Recommendations, codes of practice, etc.);

(b) the compilation of technical studies;

(c) direct assistance to governments by furnishing experts, providing fellowships, supplying equipment, drafting regulations, supplying information, etc.;

(d) assistance to national safety organisations, research centres, employers' associations, trade unions, etc., in different countries;

(e) the running of an international information centre on occupational safety and health problems.

The Governing Body, which is a tripartite body composed of government, employer and worker members, has various functions which include fixing the agenda of the International Labour Conference, and exercising general control over the activities of the Office and the advisory committees attached to it.

The International Labour Conference, which meets annually, is composed of national delegations made up of representatives of governments and of the most representative organisations of employers and workers in the country concerned. One of its functions is to discuss and adopt Conventions and Recommendations, many of which, as will be seen later, deal with matters of occupational safety and health. The Conference may also adopt resolutions calling for national or international action in this field. The member countries are required to submit Conventions adopted by the Conference to their competent authorities, with a view to ratification. Recommendations are not subject to ratification and are not binding in the same way as Conventions, but governments are bound to consider them and decide whether their provisions are acceptable or not.

The safety activities of the ILO

In the early years of the ILO, relatively little importance was attached to industrial safety. The First Session of the International Labour Conference, held at Washington in 1919, dealt, among other things, with the related field of industrial hygiene; but at that time no items dealing directly with safety were placed on the agenda. However, an important contribution was made indirectly in the form of a draft Convention fixing the minimum age for admission of children to industrial employment. Although the ILO first strictly entered the safety field in 1923, a number of Conventions and Recommendations had been adopted earlier – the White Lead (Painting) Convention, 1921 (No. 13), and the Anthrax Prevention Recommendation, 1919 (No. 3), the Lead Poisoning (Women and Children) Recommendation, 1919 (No. 4), and the White Phosphorus Recommendation, 1919 (No. 6).

In 1923, the First International Conference of Labour Statisticians made recommendations on accident rates and the classification of industrial accidents. The first safety handbook, published by the ILO in 1924, was a report on statistics of accidents due to coupling and uncoupling operations on railways. Publication of a journal called *Industrial Safety Survey* was started in 1925, and, in 1951, the title was changed to *Occupational Safety and Health*. This publication was discontinued after the foundation of the International Occupational Safety and Health Information Centre (CIS) in 1959. The abbreviated title of the Centre, CIS, is an acronym formed from its French title. The purpose of CIS is to provide information and documentation services aimed at improving occupational safety and health throughout the world. The Centre was set up as a non-profit-making organisation, under the auspices of the ILO, in collaboration with other international and national specialised bodies, and at the time of writing it is supported in this work by some 40 national centres.

In 1925, a Correspondence Committee on Accident Prevention, composed of experts from several countries, was set up to advise the ILO on measures to be taken in that field. The first safety Convention (Protection against Accidents (Dockers) Convention (No. 28)) was adopted by the International Labour Conference in 1929, and revised in 1932. Recommendations on safety (Prevention of Industrial Accidents Recommendation (No. 31), Protection against Accidents (Dockers) Reciprocity Recommendation (No. 33), and Power-driven Machinery Recommendation (No. 32)) were adopted by the Conference for the first time in 1929. A new departure was made in 1937, when a temporary committee of experts was appointed to help the Office draft safety provisions for coal mines; hitherto, the Office had relied on members of the Correspondence Committee for help of this kind. Industrial Committees began to meet in 1945. Another new step was taken in 1949, when safety provisions for industrial establishments were incorporated in the *Model code of safety regulations for industrial establishments for the guidance of governments and industry*. Supplements were added to this *Model code* in 1956 and 1959, and it is expected that a revised edition will be available during the 1980s. This code was not related to any Convention or Recommendation, and had no binding force whatsoever; it was intended solely to serve as a model for drafters of regulations on the subject. Technical assistance on safety matters was reorganised in 1951 when the United Nations Expanded Programme of Technical Assistance came into operation.

The first of a new series of documents known as codes of practice was drawn up in 1956. These codes are somewhat less formal than the *Model code*. They include the following:

— *Prevention of accidents due to fires underground in coal mines* (Geneva, 1959).

— *Prevention of accidents due to electricity underground in coal mines* (Geneva, 1959).

— *Safety and health in agricultural work* (Geneva, 1965).

— *Radiation protection in the mining and milling of radioactive ores* (Geneva, 1968).

— *Safety and health in forestry work* (Geneva, 1969).

— *Safe construction and installation of electric passenger, goods and service lifts* (Geneva, CIRA/ILO, 1972).

— *Safety and health in building and civil engineering work* (Geneva, 1972).

— *Prevention of accidents due to explosions underground in coal mines* (Geneva, 1974).

157

- *Safety and health in shipbuilding and ship repairing* (Geneva, 1974).
- *Code of safety for fishermen and fishing vessels*. Part A: *Safety and health practice for skippers and crews*. Part B: *Safety and health requirements for the construction and equipment of fishing vessels* (London, FAO/ILO/IMCO [1975]). Available from the International Maritime Organisation, London.
- *Safe construction and operation of tractors* (Geneva, 1976).
- *Protection of workers against noise and vibration in the working environment* (Geneva, 1977).
- *Safety and health in dock work* (Geneva, revised edition, 1977).
- *Accident prevention on board ship at sea and in port* (Geneva, 1978).
- *Safe design and use of chain saws* (Geneva, 1978).
- *Occupational exposure to airborne substances harmful to health* (Geneva, 1980).
- *Safety and health in the construction of fixed offshore installations in the petroleum industry* (Geneva, 1981).
- *Occupational safety and health in the iron and steel industry* (Geneva, 1983).

Shortly after this, it was decided to publish a series of supplementary guides and manuals.

- *Labour inspection: Purposes and practice* (Geneva, 1973).
- *Manuals of industrial radiation protection:*
Part I: *Convention and Recommendation concerning the protection of workers against ionising radiations* (Geneva, 1963).
Part II: *Model code of safety regulations (ionising radiations)* (Geneva, 1959).
Part III: *General guide on protection against ionising radiations* (Geneva, 1963).
Part IV: *Guide on protection against ionising radiations in industrial radiography and fluoroscopy* (Geneva, 1964).
Part V: *Guide on protection against ionising radiations in the application of luminous compounds* (Geneva, 1964).
Part VI: *Radiation protection in the mining and milling of radioactive ores* (Geneva, IAEA/ILO, 1968).
- *Medical supervision of radiation workers* (Vienna, ILO/WHO/IAEA, 1968).
- *Manual on radiation protection in hospitals and general practice:*
Volume 1: *Basic protection requirements*
Volume 2: *Unsealed sources*
Volume 3: *X-ray diagnosis*

Volume 4: *Radiation protection in dentistry* (Geneva, ILO/IAEA/WHO, 1974, 1975, 1976 and 1977). Available at WHO, Geneva.

— *Guide to the prevention and suppression of dust in mining, tunnelling and quarrying* (Geneva, 1965).
— *Guide to safety and health in forestry work* (Geneva, 1968).
— *The role of medical inspection of labour* (Geneva, 1968).
— *Guide to safety in agriculture* (Geneva, 1969).
— *Medical first aid guide for use in accidents involving dangerous goods* (London, IMCO/WHO/ILO, 1973). Available from the International Maritime Organisation, London.
— *Guide to safety and health in dock work* (Geneva, 1976).
— *Guide to health and hygiene in agricultural work* (Geneva, 1979).

In addition, the following international labour Conventions and Recommendations are of special interest:

Conventions

No. 13 White Lead (Painting), 1921
No. 18 Workmen's Compensation (Occupational Diseases), 1925
No. 27 Marking of Weight (Packages Transported by Vessels), 1929
No. 28 Protection against Accidents (Dockers), 1929
No. 32 Protection against Accidents (Dockers) (Revised), 1932
No. 42 Workmen's Compensation (Occupational Diseases) (Revised), 1934
No. 62 Safety Provisions (Building), 1937
No. 77 Medical Examination of Young Persons (Industry), 1946
No. 78 Medical Examinations of Young Persons (Non-Industrial Occupations), 1946
No. 81 Labour Inspection, 1947
No. 115 Radiation Protection, 1960
No. 119 Guarding of Machinery, 1963
No. 120 Hygiene (Commerce and Offices), 1964
No. 121 Employment Injury Benefits, 1964
No. 124 Medical Examination of Young Persons (Underground Work), 1965
No. 127 Maximum Weight, 1967
No. 129 Labour Inspection (Agriculture), 1969
No. 133 Accommodation of Crews (Supplementary Provisions), 1970
No. 134 Prevention of Accidents (Seafarers), 1970

Accident prevention

No. 136 Benzene, 1971
No. 139 Occupational Cancer, 1974
No. 148 Working Environment (Air Pollution, Noise and Vibration), 1977
No. 149 Nursing Personnel, 1977
No. 152 Occupational Safety and Health (Dock Work), 1979
No. 155 Occupational Safety and Health, 1981

Recommendations

No. 3 Anthrax Prevention, 1919
No. 4 Lead Poisoning (Women and Children), 1919
No. 6 White Phosphorus, 1919
No. 31 Prevention of Industrial Accidents, 1929
No. 32 Power-driven Machinery, 1929
No. 33 Protection against Accidents (Dockers) Reciprocity, 1929
No. 34 Protection against Accidents (Dockers) Consultation of Organisations, 1929
No. 40 Protection against Accidents (Dockers) Reciprocity, 1932
No. 53 Safety Provisions (Building), 1937
No. 54 Inspection (Building), 1937
No. 55 Co-operation in Accident Prevention (Building), 1937
No. 79 Medical Examination of Young Persons, 1946
No. 81 Labour Inspection, 1947
No. 82 Labour Inspection (Mining and Transport), 1947
No. 97 Protection of Workers' Health, 1953
No. 112 Occupational Health Services, 1959
No. 114 Radiation Protection, 1960
No. 118 Guarding of Machinery, 1963
No. 120 Hygiene (Commerce and Offices), 1964
No. 121 Employment Injury Benefits, 1964
No. 128 Maximum Weight, 1967
No. 133 Labour Inspection (Agriculture), 1969
No. 140 Crew Accommodation (Air Conditioning), 1970
No. 141 Crew Accommodation (Noise Control), 1970
No. 142 Prevention of Accidents (Seafarers), 1970
No. 144 Benzene, 1971
No. 147 Occupational Cancer, 1974

No. 156   Working Environment (Air Pollution, Noise and Vibration), 1977

No. 157   Nursing Personnel, 1977

No. 160   Occupational Safety and Health (Dock Work), 1979

No. 164   Occupational Safety and Health, 1981

A full list of all the relevant publications is given at the back of this manual.

Of special interest are the latest Convention and Recommendation concerning safety and health and the working environment, which were adopted in 1981. For the first time, the whole question of the prevention of occupational hazards and the improvement of the working environment was looked at in its entirety. The Convention applies to all branches of economic activity, including the public service. At the national level, it lays down that there must be a tripartite approach to formulating, implementing and periodically reviewing a coherent national policy on occupational safety, health and the working environment. Such a policy should aim at preventing accidents and injury to health arising out of, linked with, or occurring in the course of work, by minimising, so far as is reasonably practicable, the causes of hazards inherent in the working environment. Article 5 of this Convention takes into account the following main spheres of action:

(a) design, testing, choice, substitution, installation, arrangement, use and maintenance of the material elements of work (workplaces, working environment, tools, machinery and equipment, chemical, physical and biological substances and agents, work processes);

(b) relationships between the material elements of work and the persons who carry out or supervise the work, and adaptation of machinery, equipment, working time, organisation of work and work processes to the physical and mental capacities of the workers;

(c) training, including necessary further training, qualifications and motivations of persons involved, in one capacity or another, in the achievement of adequate levels of safety and health;

(d) communication and co-operation at the levels of the working group and the undertaking and at all other appropriate levels up to and including the national level;

(e) the protection of workers and their representatives from disciplinary measures as a result of actions properly taken by them in conformity with the coherent national policy referred to above.

Other salient points are that workers should be given adequate information and appropriate training, or should be able to inquire into, and be consulted by the employer on, all aspects of occupational safety and health. Workers should, moreover, be able to report forthwith any situation which could present an imminent and serious danger. The employer cannot require workers to return to a work situation where such danger persists before having taken remedial action. Workers who have removed themselves from such a work situation shall be protected from undue consequences.

The Convention requires employers to ensure that workplaces, machinery, equipment and processes under their control, as well as chemical, physical and biological agents and substances used within the undertaking, are safe and without risk to health. Employers must also provide adequate protective clothing and protective equipment to prevent risk of accidents or of adverse effects on health.

The Recommendation specifies technical fields of preventive action, taking into account the diversity of branches of activity, and of types of work, and the principle of giving priority to eliminating hazards at their source.

Readers are advised to look at this Convention and Recommendation to obtain a full understanding of them and of their possible implications.

INTERNATIONAL TECHNICAL CO-OPERATION

Technical co-operation programmes, such as those sponsored during recent years by the United Nations, the organisations belonging to the United Nations family, including the International Labour Organisation, and many individual countries throughout the world, have been playing a prominent part in promoting occupational safety and health in developing countries. Essentially, these activities are designed to apply the rich and varied experiences of the more industrialised countries to those that are in the process of industrialisation, and to adapt them according to local factors and circumstances. Through its various branches, the ILO spends a considerable amount of time and effort in the search for solutions to old problems, and in meeting new challenges brought about by changing technology.

A wide variety of activities, many of which have a direct bearing on the promotion and strengthening of the world network of the occupational safety and health movement, are carried out.

Associated in this work are employers' and workers' organisations, and official, semi-public and private organisations and societies, all of which can, and do, make a significant contribution to the battle against accidents and diseases associated with employment. In this way, it is possible to carry services, techniques and procedures straight to shop-floor level where they can be built into the normal processes and operations of production. Information and training in workers' protection are increasingly being integrated into the Organisation's numerous other projects in the field.

This effort towards international technical co-operation is largely financed by the United Nations Development Programme (UNDP). Between 1950 and 1978, the UNDP entrusted the ILO with carrying out more than 250 major projects.

Under the ILO's regular budget, or under the United Nations Development Programme, experts have been sent to over 50 countries in all parts of the world to supply advice on the organisation of occupational safety and health services, factory inspection and safety in mines. The general procedure is for an expert, or a team of experts, to make a survey of industrial conditions, legislation and administrative facilities, and to recommend suitable action in each field.

The ILO is also currently devoting a considerable part of its resources from the United Nations Development Programme to meeting requests for assistance in setting up institutes to help to adapt work to people, and people to their work. These institutes cover a wide variety of scientific and technical fields (e.g. nutrition, pollution control in the working environment, psychology, physiology and sociology, the broad field of ergonomics) which contribute to relieving mental and physical stress, and to making work less arduous, more productive and more rewarding. Their main function is to formulate appropriate measures, taking into account the particular conditions of the country, to give direct assistance to industry in applying such measures in specific situations, to give all those responsible for occupational safety and health an opportunity for further training, and to carry out research into occupational safety and health problems.

Particular attention is given not only to providing support materials for education and training, but also to ensuring up-to-date information and operational research for better planning and implementation of technical co-operation activities. This line of action calls for the organisation of numerous symposia, seminars and meetings, the proceedings and conclusions of which become valuable instruments in training courses in general, or on specific safety and health subjects. The range of such courses extends from protection against ionising radiations, and dust prevention, to factory inspection techniques, and the safe use of equipment and chemicals in agriculture.

In brief, technical co-operation, to be effective, must have recourse to the many techniques available in accident prevention and related work, and must continue to develop new approaches and methods to keep up with rapid industrialisation and the introduction of new production processes and operations.

PIACT

In 1974, the International Labour Conference adopted a resolution in which it was stressed that "the improvement of the working environment should be considered a global problem in which the various factors affecting the physical and mental well-being of the worker are inter-related". Subsequently, in 1975, the Conference unanimously adopted a resolution reaffirming that the improvement of the working conditions and environment, and the well-being of workers, remains the primary and permanent aim of the ILO. Accordingly, an International Programme for the Improvement of Working Conditions and Environment, known as PIACT from its French acronym, was launched in response to the specific request of the 1975 Session of the International Labour Conference. The main objectives of PIACT are as follows:

(a) to encourage member States to set definite objectives designed to improve working conditions and environment, particularly by promoting the effective application of international labour standards setting forth such objectives;

(b) to propose, if need be, the adoption of new international standards to define new objectives; and

(c) to provide governments, employers' and workers' organisations and research and training institutions with the necessary assistance for the preparation and implementation of programmes for the improvement of working conditions and environment corresponding to their potential.

QUESTIONS

1. Give reasons why international action should be taken to promote occupational safety.

2. Mention some of the activities developed by the ILO in the occupational safety field.

TRADE UNIONS, WORKERS AND INDUSTRIAL SAFETY

14

TRADE UNIONS

Workers' unions have a very direct interest in promoting safety, for in nearly all cases it is workers who are killed and injured in accidents.

There are many ways in which unions can contribute to safety, and in fact some of them have taken direct action to ensure high standards of safety.

One important measure which unions can take is to have safety provisions included in collective agreements. Most of the agreements involved contain only general provisions concerning safety and health, although in particularly hazardous occupations, and where the safety of the general public is involved, they may include detailed rules and requirements. Generally, any differences between the union and the employers regarding safety and welfare may be referred to the regular grievance procedure, or other joint machinery, for adjustment. Under some agreements, such matters are referred to a special safety committee, which may be a joint management-union committee or one composed solely of union members. Others require the employer to maintain first-aid facilities in the plant and injured employees to report accidents. Since the employer is usually required to report all accidents under workmen's compensation legislation, most agreements do not contain clauses on the subject.

Co-operation between unions and employers takes a number of forms. Some unions provide for equal representation with the management on safety committees, and the union representatives are appointed or elected directly. Others list a certain number of names of members, and the management makes the final selection of a predetermined number (e.g. three names from a list of ten). In some cases, the union nominees have to be approved by their supervisors.

Sweden is a country in which co-operation between the unions and the other main parties interested in safety – the Government and the employers – is highly organised. The unions take the view that safety cannot be ensured by legislation alone, and that the latter must be supplemented by co-operation at all levels, from national to local.

In some countries, trade unions actually assist the labour inspection services. Sometimes, the workers appoint officials to the inspection service for a specific period, and these officials help with the enforcement of regulations dealing with working conditions. In other cases, trade union leaders are recruited as inspecting officials.

It is good that trade unions should strive to secure better wages, working hours, holidays, social security benefits and other means of improving the living conditions of their members, but it might also be said that the first duty of trade unions should be to do what they can to keep their members alive and intact. Good living conditions are of no use to a dead man; a widow's pension cannot replace a dead husband; and no amount of compensation will restore sight to a blind man.

In conclusion, one sort of union activity which is not conducive to safety should be mentioned. Cases have occurred in which trade unions have demanded additional payment for their members (known as "danger money" or "dirt money") for work considered to be particularly dangerous. Safety is not promoted in this way, for the work, and the risks, remain the same, whatever the rate of payment. The idea of danger money has proved to be extremely harmful, for all too often unions have attempted to get work classified as dangerous rather than to have the danger removed.

WORKERS' ATTITUDES

In the preceding lessons, the different aspects of industrial safety have been dealt with. The economic and financial consequences of accidents have been described, and it has been seen that, although economic considerations are important, the main reason for safety activities is the moral obligation to protect the workers from physical danger. Various kinds of preventive measures, and the responsibilities of the various parties concerned, have been examined from this point of view. We may well now ask what the workers themselves think about all this, and what they do, and should do, in the interest of their own safety and that of their fellow workers. This question is difficult to answer, because there are fundamental differences in the living conditions of workers, their attitudes to their work, and their ways of thinking, in different parts of the world.

It may be assumed that workers want decent working conditions. In many countries, they have shown a dislike of working conditions regulated solely by the employer's patronage or favours. They want conditions conforming to their conceptions of rightness, justice and equity. Where this is the case, they may well feel that such conditions have usually been won as the outcome of long social struggles involving strikes and lock-outs, prosecutions and persecutions. In these struggles, many years of joint action by the workers have brought forth feelings of solidarity, and of a deep distrust of the employers.

Today, in some countries, workers have extensive rights which are generally accepted by the employers, and this distrust has more or less given way to feelings of mutual responsibility. In others, trade unions hardly exist, and workers are not organised, except perhaps in very small associations, embracing, for instance, a single factory. In technically advanced countries, we find trade unions which are guided by people with an expert knowledge of labour problems, and which employ technical advisers; but in developing countries, such unions are few.

Some differences in the attitude of workers may be traced to religious influences. It makes a great difference, for instance, whether a religion stresses mutual responsibility or teaches a fatalistic acceptance of things as they are.

Many workers go to the factory only to earn money for themselves and their family; they have no special interest in their jobs, and they may work in the factory for years without any bond forming between their inner selves and their work. They have little love or trust for their employer, and they are not likely to take kindly to advice and exhortations on safety or anything else.

Workers' attitudes to industrial safety may, in fact, depend on a whole array of factors, ranging from their social and religious background to their own circumstances and character. It must be said that workers themselves are usually not the driving force in accident prevention activities. Even in technically advanced countries, where workers are relatively well off, tremendous efforts are required to make workers safety-minded. This seems to show that workers are seldom spontaneously interested in safety, even though their lives may be at stake. Individual workers have no doubt made great contributions to safety as members of safety committees, or in some other capacity, but, on the whole, improvements in safety cannot be said to have originated from workers. This can perhaps be partly explained by the fact that, almost everywhere, the law makes the employer responsible for establishing and maintaining safe working conditions; but the probable explanation is that workers are more interested in questions of wages, hours of work, holidays, compensation, the closed shop, etc., than in questions of safety. It is also true that workers are accustomed to their working environment, and to its risks. Underestimation of these risks, and a false feeling of immunity from them, tend to make workers relatively indifferent towards safety matters.

In developing countries, workers are often ignorant of the risks to which they are exposed. Many of them are peasants who are recruited from the villages, and who remain only for a short period in industry. They are often illiterate. It is therefore not surprising that they show little interest in accident prevention.

If it is generally true that workers have a very limited interest in accident prevention, we may wonder if the situation is likely to improve. As all the efforts that have been made for very many years to arouse and intensify workers' interest have had only a very limited success, it cannot be expected that the situation will improve rapidly; but there can be little doubt that safety education, in its various forms, will be beneficial in the long run. It would, anyway, be unthinkable to abandon the struggle.

The solidarity constantly shown by workers in times of trouble should manifest itself in the field by safety. All workers have a duty and a responsibility to protect their fellow workers from accidents. Workers must not think "Am I my brother's keeper?"; they must not stand aside when they see a fellow worker taking serious risks, for this would be a grave breach of duty.

It was stated earlier that the accident frequency rate among young workers is relatively high. This is due sometimes to insufficient knowledge of risks, but more often to unnecessary actions motivated by a desire to do more than is absolutely necessary, to satisfy curiosity, or to indulge in horseplay. Here, the older and more experienced workers have a definite responsibility to give guidance and supervision. The number of accidents in which young persons are involved might well be reduced considerably if this responsibility were more generally exercised.

QUESTIONS

An excellent method by which workers can explore the field of accident prevention further, either privately or in groups, is through research into their own, or their union's, situation. To stimulate further study, the questions below have been designed to cover a wider field than the matter contained in this lesson. They may provide a basis for the review of specific ideas brought out in previous lessons; they may also suggest certain subjects for group discussion. They are not intended to provide a framework for an exhaustive survey of the whole subject; but they will — with adaptations and complements — help readers to discover for themselves the practical implications of the knowledge acquired from the present course, and perhaps also to formulate the elements of a constructive union policy on safety.

The safety situation

1. If your union had to formulate or review its policy with respect to safety matters in a factory, what technical information and statistical data

should it first obtain to determine that policy and justify subsequent claims or action? (Some useful lines of investigation will be found in Lessons 3 and 11, in particular.)

Workers' attitudes

2. Are the workers in your factory interested in promoting safety measures? How do you recognise this interest? What do they do in the interest of safety?

3. What factors (for instance, special dangers of the trade and frequency of accidents, education and propaganda programmes by the union, by the employer, by the labour inspection services) give rise to this attitude? In the last two cases, did the union co-operate? And if so, how? (See Lesson 8, in particular.)

4. If there is lack of interest for an active safety attitude, what are its causes? (Examples: ignorance of risks, underestimation of risks, lack of solidarity between workers.) What remedies are there to this situation? In particular, what can the union do?

The trade unions

5. What safety provisions appear in the collective agreement covering your factory?

6. Are there first-aid facilities in the plant?

7. Are injured employees requested or required to report accidents?

8. What is the procedure for dealing with differences between the union and the employer on safety questions?

9. Is there a plant safety committee? Is it a union committee or a joint body? In the latter case, how are the worker members appointed? Do they receive specialised training for this job from the union, or otherwise?

10. In what ways (e.g. literature, campaigns, exhibitions, films) is safety promoted? How does the union co-operate in this educational effort?

11. Is there a joint or tripartite safety body at the industrial or national level?

PUBLICATIONS OF THE ILO

MODEL CODES

Model code of safety regulations for industrial establishments for the guidance of governments and industry (Geneva, 3rd impression, 1962), 523 pp.

Model code of safety regulations in underground work in coal mines (Geneva, 1950), 102 pp.

CODES OF PRACTICE

Prevention of accidents due to fires underground in coal mines (Geneva, 1959), 48 pp.

Prevention of accidents due to electricity underground in coal mines (Geneva, 1959), 54 pp.

Safety and health in agricultural work (Geneva, 4th impression, 1983), 132 pp.

Safe construction and installation of electric passenger, goods and service lifts (Geneva, 1972), 108 pp.

Safety and health in building and civil engineering work (Geneva, 3rd impression, 1985), 386 pp.

Prevention of accidents due to explosions underground in coal mines (Geneva, 1974), 37 pp.

Safety and health in shipbuilding and ship repairing (Geneva, 3rd impression, 1984), 260 pp.

Safe construction and operation of tractors (Geneva, 1976), 39 pp.

Protection of workers against noise and vibration in the working environment (Geneva, 3rd impression, 1984), 74 pp.

Safety and health in dock work (Geneva, revised edition, 1977, 3rd impression, 1984), 221 pp.

Accident prevention on board ship at sea and in port (Geneva, 4th impression, 1984), 188 pp.

Safe design and use of chain saws (Geneva, 1978), 71 pp.

Code of safety for fishermen and fishing vessels

   Part A: *Safety and health practice for skippers and crews*

   Part B: *Safety and health requirements for the construction and equipment of fishing vessels*

(London, FAO/ILO/IMCO [1975]). Available from the International Maritime Organisation, London.

Occupational exposure to airborne substances harmful to health (Geneva, 2nd impression, 1985), 40 pp.

Safety and health in the construction of fixed off-shore installations in the petroleum industry (Geneva, 1981), 132 pp.

Occupational safety and health in the iron and steel industry (Geneva, 1983), 343 pp.

Safety in the use of asbestos (Geneva, 1984), 116 pp.

Safety and health in coal mines (in preparation).

ENCYCLOPAEDIA

Encyclopaedia of occupational health and safety (3rd (revised) edition, 1983). Two volumes.

GUIDES AND MANUALS

Labour inspection: Purposes and practice (Geneva, 3rd impression, 1985), 244 pp.

Manual of industrial radiation protection:

Part I: *Convention and Recommendation*[2] (Geneva, 1963), 24 pp.

Part II: *Model code of safety regulations (ionising radiations),*[3] (Geneva, 2nd impression, 1965) (out of print), 54 pp.

Part III: *General guide on protection against ionising radiations* (Geneva, 1963), 95 pp. Illustrated.

Part IV: *Guide on protection against ionising radiations in industrial radiography and fluoroscopy* (Geneva, 2nd impression, 1965), 62 pp. Illustrated.

Part V: *Guide on protection against ionising radiations in the application of luminous compounds* (Geneva, 1964), 48 pp. Illustrated.

Part VI: *Radiation protection in the mining and milling of radioactive ores*[4]

Medical supervision of radiation workers (Vienna, ILO/WHO/IAEA, 1968), 118 pp.

Manual on radiation protection in hospitals and general practice:

Volume 1: *Basic protection requirements*

Volume 2: *Unsealed sources*

Volume 3: *X-ray diagnosis*

Volume 4: *Radiation protection in dentistry*

(Geneva, ILO/IAEA/WHO, 1974, 1974, 1975, 1977).

Guide to the prevention and suppression of dust in mining, tunnelling and quarrying (Geneva, 1965), 421 pp. Illustrated.

The role of medical inspection of labour (Geneva, 1968), 111 pp.

Guide to safety in agriculture (Geneva, 1969), 247 pp.

Medical first aid guide for use in accidents involving dangerous goods (London, IMCO/WHO/ILO, 1973). Available from the International Maritime Organisation, London.

Guide to safety and health in dock work (Geneva, 1976), 287 pp.
Guide to health and hygiene in agricultural work (Geneva, 1979), 309 pp.

OCCUPATIONAL SAFETY AND HEALTH SERIES[5]

7. *Organisation of occupational health services in developing countries* (Geneva, 2nd impression, 1976), 187 pp.
9. *Dust sampling in mines* (Geneva, 2nd impression, 1972), 88 pp.
16. *International directory of occupational safety and health institutions* (Geneva, 3rd edition, 1977), 520 pp.
17. *International catalogue of occupational safety and health films* (Geneva, 1969), 557 pp.
22. *ILO guide-lines for the use of international classification of radiographs of pneumoconioses* (Geneva, revised edition, 1980, 4th impression, 1984), 24 pp.
26. *Occupational health problems of young workers* (Geneva, 3rd impression, 1983), 143 pp.
27. *Safety and health in shipbuilding and ship repairing* (Geneva, 1972), 236 pp.
28. *Safe construction and installation of escalators* (Geneva, 1976), 29 pp.
29. *Médecine du travail, protection de la maternité et santé de la famille* (Geneva, 1975), 58 pp. French only.
31. *Organisation of family planning in occupational health services* (Geneva, 1976), 47 pp.
32. *Radiation protection in mining and milling of uranium and thorium* (Geneva, 1976), 346 pp.
33. *Noise and vibration in the working environment* (Geneva, 1976), 131 pp.
34. *Migrant workers — Occupational safety and health* (Geneva, 1977), 82 pp.
38. *Safe use of pesticides* (Geneva, 3rd impression, 1985), 42 pp.
40. *5th international report on the prevention and suppression of dust in mining, tunnelling and quarrying* (Geneva, 2nd impression, 1980), 106 pp.
41. *Safety and health of migrant workers — International symposium* (Geneva, 2nd impression, 1983), 337 pp.
42. *Building work — A compendium of occupational safety and health practice* (Geneva, 1979), 212 pp.
43. *Optimisation of the working environment — New trends* (Geneva, 1979), 429 pp.
45. *Civil engineering work — A compendium of occupational safety practice* (Geneva, 1981), 153 pp.
46. *Prevention of occupational cancer — International symposium* (Geneva, 1982), 658 pp.
47. *Education and training policies in occupational safety and health and ergonomics — International symposium* (Geneva, 1982), 389 pp.
48. *6th international report on the prevention and suppression of dust in mining, tunnelling and quarrying* (Geneva, 1982), 152 pp.
49. *Dermatoses et professions* (Geneva, 1983), French only.
50. *Human stress, work and job satisfaction* (Geneva, 1983), 72 pp.
51. *Stress in industry: Causes, effects and prevention* (Geneva, 1984), 70 pp.
52. *Success with occupational safety programmes* (Geneva, 1984), 148 pp.

Accident prevention

53. *Occupational hazards from non-ionising electromagnetic radiations* (Geneva, 1985), 133 pp.
54. *The cost of occupational accidents and diseases* (in preparation).
55. *The provisions of the basic safety standards for radiation protection relevant to the protection of workers against ionising radiation* (Geneva, 1985), 23 pp.
56. *Psycho-social factors at work: Recognition and control* (in preparation).

VISUAL AIDS

Occupational safety and health films available on loan at the ILO. SHC/Fm.1.(rev.) (Geneva, 1979), 60 pp.

Learning to make work safer and healthier (28 minutes, colour, English, 16 mm, 1980). A film available from the Workers' Education Branch of the ILO.

An investment in safety (37 minutes, colour, English, German, 16 mm, 1976). A film available from the Workers' Education Branch of the ILO.

Making work more human (50 frames, double-frame format, 35 mm, colour). A film strip, accompanied by a sound cassette and text booklet, available from the Workers' Education Branch of the ILO.

International classification of radiographs of pneumoconioses (Geneva, revised, 1980). Set of 22 standard films.

COMPILATION OF ACCIDENT STATISTICS

Statistics of industrial injuries. 1970. Doc. D 17.1970/x.CIST/II/SAT. 56 pp.
Year Book of Labour Statistics (Geneva, 1985), 962 pp.

PUBLICATIONS OF THE INTERNATIONAL OCCUPATIONAL SAFETY AND HEALTH INFORMATION CENTRE (CIS)[6]

CIS Information Sheets

CIS Information Sheets have now been replaced by Information Notes. The following Information Sheets were still available at time of going to press:

7. *Circular saws,* 50 pp. Illustrated.
10. *Ergonomics of machine guarding,* 18 pp. Illustrated.
11. *Artificial lighting in factory and office,* 61 pp. Illustrated.
12. *Ladders,* 74 pp. Illustrated.
16. *Safety in the construction of reinforced-concrete floors,* 26 pp. Illustrated.
18. *Asbestos-cement roofs,* 13 pp. Illustrated.
19. *Fifty years of international collaboration in occupational safety and health,* 72 pp.
20. *Diving,* 31 pp. Illustrated.

CIS bibliographies

10. *Vinyl chloride,* 12 pp.

11. *Occupational cancer*, 64 pp.

12. *Forestry*, 28 pp.

13. *Asbestos*, 60 pp.

14. *Noise*, 52 pp.

15. *Work with display units*, 12 pp.

16. *Electromagnetic radiation*, 14 pp.

17. *Vibration*, 32 pp.

18. *Printing industry*, 12 pp.

19. *Pneumoconiosis*, 60 pp.

20. *Economic aspects*, 12 pp.

21. *Asbestos. No. 2*, 37 pp.

22. *Health services*, 12 pp.

23. *Transfer of technology*, 12 pp.

24. *Coal-mining*, 52 pp.

CIS Thesaurus

More than 7,000 descriptors and alphabetical index with 16,000 entries.

Notes

[1] In 1956, the ILO revised the regulations on the textile industry and welding operations, in 1959 those concerning ionising radiations (see under "Guides and manuals") and in 1970 those concerning elevators and built-in hoists (see above, under "Codes of practice").

[2] Contains the text of the Radiation Protection Convention and Recommendation, 1960 (No. 115 and No. 114).

[3] Revised text of Section 2 of Chapter XI of the *Model code of safety regulations for industrial establishments for the guidance of governments and industry*, mentioned at the beginning of this list.

[4] Published jointly with the International Atomic Energy Agency in Vienna.

[5] Information on distribution may be obtained direct from the Occupational Safety and Health Branch, International Labour Office, CH-1211 Geneva 22 (Switzerland).

[6] Information on distribution may be obtained directly from the International Occupational Safety and Health Information Centre (CIS), International Labour Office, CH-1211 Geneva 22 (Switzerland).

Other ILO publications

Introduction to working conditions and environment
Edited by J.-M. Clerc

The state of working conditions and the working environment in many parts of the world can only be described as alarming: for it is no secret that the lives, health and welfare of millions of workers are at stake during each moment of the working day. For many years, the need has been felt for a wide-ranging introductory volume embracing the main aspects of occupational safety and health and general conditions of work. This book has been designed to meet that need. By attempting to disseminate knowledge and information as widely as possible, both through this book and through its International Programme for the Improvement of Working Conditions and Environment, the ILO is expressing its clear conviction that this is a task in which all must participate – not only those who are experts in the subject, but governments, employers and workers as well.

ISBN 92-2-105124-2 (Hard cover); ISBN 92-2-105125-0 (Limp cover)

Safety and health practices of multinational enterprises

Technological change and the emergence of new products and production processes in the industrialised countries, together with growing industrialisation in Third World countries, have brought the question of occupational safety and health into sharper focus. As major employers and as innovators of advanced technology in many parts of the world, multinational enterprises have a key role to play in the development of protective measures for those of their workers who are exposed to potentially hazardous processes and substances.

In this monograph the occupational safety and health standards prevailing in eight selected MNEs from a cross-section of industry in seven countries (Federal Republic of Germany, Mexico, Netherlands, Nigeria, Switzerland, United Kingdom, United States) have been investigated, through in-depth interviews with management and workers in these MNEs and through on-site inspections at 12 of their subsidiaries. An attempt has also been made to ascertain how the pertinent information on this subject is transferred from the parent companies to their various subsidiaries.

ISBN 92-2-103742-8

Wages. A workers' education manual

The more trade union leaders and members know about what determines wage levels, the causes of differences between wages in various occupations, industries and regions, and the methods of wage payment (time-rates, piece-rates and bonuses); the more effective will their action be in securing improvements, removing anomalies that are unjustifiable and establishing proper wage structures.

This volume, consisting of 16 lessons, has been written mainly for use in study courses attended by trade union members and other workers so that they may gain a clear understanding of the issues involved. This is not a treatise on economics but a practical tool that can be adapted to the widely varying circumstances characteristic of workers' education.

The manual covers such matters as the bases for fixing wages, systems of payment by results, fringe benefits and profit-sharing schemes, job evaluation, wage-fixing methods, women's wages, the protection of wages, wage theories, problems of national incomes policy and international wage problems and international labour standards on wages.

There are three appendices containing reference texts, a glossary and suggestions for further reading.

This third edition of the manual has been revised and updated in the light of developments and experience in the years since the publication of the original edition.

ISBN 92-2-102961-1